Your Teen's *Miraculous* BRAIN

Your Teen's *Miraculous* BRAIN

Eight Drug Free Essentials for Overcoming Teen Mental Illness

Dr. Nina Farley-Bates

NEW YORK

LONDON • NASHVILLE • MELBOURNE • VANCOUVER

Your Teen's *Miraculous* BRAIN
Eight Drug Free Essentials for Overcoming Teen Mental Illness

Published in New York, New York, by Morgan James Publishing in partnership with Difference Press. Morgan James is a trademark of Morgan James, LLC.
www.MorganJamesPublishing.com

ISBN 978-1-64279-359-8 paperback
ISBN 978-1-64279-360-4 eBook
Library of Congress Control Number: 2018913681

Interior Design by:
Bonnie Bushman
The Whole Caboodle Graphic Design

Dedicated to my four children as they discover with every passion in their hearts, with all the energy in their being, and every thought within them that it is possible to love others as they love themselves.

Contents

Introduction

If you have felt hopeless, hold on! Wonderful changes are going to happen in your life as you begin to live it on purpose.
—Rick Warren

Advice abounds for parents of teens, but reliable, helpful information that works for parents worried about their teens is scarce. If you haven't already done so, just jump on the internet and query a few hundred sites. What you'll notice first are those intrusive pop-up advertisements or pharmaceutical-sponsored sites pushing prescription medications or teen treatment programs. After you scroll down further, you'll find short articles, authored by professional writers who often have little understanding of teen mental illness or what parents need to know to help them succeed. Sadly, buried under all those poorly informed parenting tips, you'll find parents equally as discouraged as you with mentally ill teens who, after searching the internet themselves, are still frantically asking the same questions you are on behalf of your teen.

According to the National Alliance on Mental Illness (NAMI), many psychiatric illnesses, including major depression, anxiety disorders, and psychotic disorders, first appear between the ages of 14 and 25, with other conditions more commonly diagnosed in much younger children, such as attention deficit/hyperactivity disorder (ADHD/ADD) or oppositional defiant disorder. In a world where information is everywhere for the downloading, you may be painfully aware that what attracts more hits on the internet gets more teens into expensive treatment programs or sells more parents on another pharmaceutical, and, in the end, fails to provide the information you need to help your tween, teen or young adult.

Your teen may already be under the care of a psychiatrist or psychotherapist, but you have seen little improvement, and when you question that lack of progress, your teen's physician hands you another script or your teen's therapist suggests perhaps it's really something you lack in your parenting that's at the root of your teen's problem.

If you're worried that your teen's flareups of despair, worry, or rage are getting worse and becoming more frequent. You may be concerned that whatever is going on with your teen involves more than typical teenage behavior. Your worst fear may be that something more serious is wrong that may result in drug abuse or worse—suicide or even homicide.

What's most confusing is that your child may have had a sheltered life up until their teen years; while all around you other couples were divorcing, you and your spouse stuck it out, so your child benefited from as stable and as happy a home as you could provide. Now, as evidenced by your teen's current behavior, all that effort to make a perfect loving home wasn't good enough, and your teen announces one day she's been so depressed she's thinking about killing herself. If you have a teen who has ever opened up to you like that, count yourself among the blessed—most children never disclose these feelings to their parents and the parents find out the hard way after an attempt or, worse, a successful suicide.

Maybe you can relate to your teen's despair, if your own childhood was something you don't want your teen to endure because you were forced to witness their vicious cycles of substance use, divorce, or domestic violence. You may have suffered physical, verbal, or sexual abuse or neglect at the hands of your parents. Any combination of these adverse childhood experiences for you adds up to the desperate loneliness you felt as a child. Still, to the best of your ability, you protected your child from all that happened to you, so what's going wrong with your teen?

Maybe your childhood wasn't all that bad; it's just that you essentially raised yourself as a kid. You were one of those latchkey kids. So, just about the time a new key opened the door to your own home, you were still growing up yourself when your child came along. Now that your child has reached adolescence, you can see clearly where all that angry self-sabotaging behavior in your teen comes from. It's much like looking in a mirror of your own child self. You realize that your teen is just like you, with a seemingly harder-to-deal-with attitude. Your determination mountain with your teen is that they will not flounder and have to endure the pain of the bad choices that you made. You just can't bear the anguish of your child graduating from the same school of hard knocks as you did.

Whatever went wrong in your childhood, you've got to give your parents some credit—at least they showed you everything not to do with your children. But just doing the opposite of your parents hasn't inoculated your teen from mental illness. It brings you little comfort to see parents of children with no identifiable past trauma who also struggle with their teens. You're wondering if, like them, your child might also be diagnosed with bipolar disorder, ADHD/ADD, anxiety, depression, or panic disorder.

You might be one of those extraordinarily wonderful parents who adopted an older child who had been prenatally drug exposed, neglected or abused by their biological parents during their infancy. With all the best intentions as a step-parent, you may have welcomed into your new marriage a special needs child that came with many challenges you had not anticipated. Now that she's a tween, you are feeling all but extraordinary. It took time, dedication, and self-sacrifice on your part to get your adopted child on a path to overcoming early exposure to childhood developmental trauma. Now, all that effort to get your child into therapy and on medication seems pointless. You're paying the price for what a lack of nurturing did to your child while he was waiting for you to bring him home from foster care. Despite your best efforts to correct your child's early failed parenting, your child's trauma is coming back with a vengeance and taking it all out on you.

You may even be an older sibling who has stepped into the role of a mom or dad with deadbeat parents of your own who left you to care for your younger siblings. Grandparents, step-parents, aunts, uncles, foster parents—we might all be coming from different places, but one thing remains: if wishes were answers or cures, your teen would be healthy, happy, whole and healed by now.

Maybe your tween, teen, or transition age youth seems to suffer from all the worst I've just mentioned rolled into one big headache and heartache with your name on it. You might be fearful as the days go by about how this is going to end for your child. You know you can't always be there to bail them out. But what really takes your breath away is all the bad advice and criticism you get about the child you care so much for. All the judgments against you as a parent just add to a sneaking suspicion that maybe you are the cause of your child's mental illness.

Mental Health in Church

For those of you who have felt unwelcomed in church because you and your child are struggling with mental illness, it's important for you to know you are not alone. In his book, *Mental Health and the Church*, physician and author Stephen Grcevich identifies the disconnect many people feel between the vastly divergent worlds we negotiate between our personal and professional lives and, for those in mental health, between the helping professions in mental health and the church.

As a psychiatrist, Dr. Grcevich finds that those families with a mentally ill child or adolescent whom he treats at Northeast Ohio Medical University are far less likely than other families in his community to be actively involved in community churches. Dr. Grcevich calls this "a tragic departure from Jesus' plan for his church." I see this as well, both in my private practice and my integrated behavioral health practice in my rural community's primary care medical clinic.

We can't afford to continue divesting the church from the support systems that improve the quality of life for mental health patients and us all. Together, the church and the mental health professions are both vital to maintaining healthy and safe communities for our teens and ourselves. Dr. Grcevich urges us to confront the following statistics:

- 8–12 percent of teens experience anxiety disorders; fewer than one in five receive any treatment—psychotherapy or medication.
- 16.1 million adults (6.7 percent of all US adults) experienced an episode of major depression during 2015.
- Suicide is the second leading cause of death in the United States (following unintentional injury) among individual ages 15 to 34.

- Suicide is the third leading cause of death among young people between the ages of 10 and 24, with a rising trend among even younger children who are victims of bullying even as young as five.
- To date, more than 187,000 students have been exposed to gun violence at school since the Columbine High School massacre occurred April 20, 1999.

Whether we're churchgoers or not, these grim statistics support what we all must face as parents and helping professionals, from medical to religious and from research institutions to private practices. We can no longer stand maintaining our opposing views and ignoring the power of our unity at least on one point: Our youth need support beyond the secular, religious, scientific, or political walls we've erected against each other.

Whatever got you here in this moment, no matter what others are saying about your parenting, you probably did little to earn these bad feelings you now wear with guilt and shame, but here you are, feeling alone and with little support, right along with your failure-to-launch child. Neither your parents nor parenting experts have answers you've been able to use—and that puts your feet back pounding pavement and searching for answers.

Empowered for Change

What has helped me as a mental health counselor and parent of four children is picking up the tools daily that worked and putting down the ones that didn't. It makes it sound so simple, doesn't it? In a way it is simple, because it's about a childlike reliance on the power to change. For me, connecting to that power as a parent has been all about becoming more like Christ. As Rick Warren of Saddleback Church says, "Becoming like Christ is a long, slow process of growth."

Similarly, growing up into more powerful parenting is a learn-as-you-grow kind of process, too. A process that starts with pushing through fear and finding your faith. Faith in the resiliency of the human brain, body and mind. Faith in the power greater than all the love and worry you could ever marshal on your own for your child. Faith that God can and will do for your teen what has become impossible for you to do alone.

Picking up these powerful tools is much like God commanding Moses, "Pick it up by the tail!" That "it" was a big old ugly snake—not something I'd be inclined to pick up on my most daring days. Assuming he didn't care much for snakes himself, Moses pushed through his own fear and with reckless abandon picked up that big scary thing exactly in the place God commanded him to, even though in the natural world, he knew that would give that snake the precise advantage needed to strike and kill him (Exod. 4:4). Picking up what looked like the most improbable tool, in what seemed like the least sensible way for Moses became his most powerful solution. That most unlikely solution became his shepherd's staff for Moses to lead his people safely to the promised land.

In their lifestyle health book *The Daniel Plan*, Pastor Rick Warren, brain health psychiatrist Daniel Amen, and wellness physician Mark Hyman assert that while most self-help books give great advice, few offer real power to change. These authors say that self-help books "tell you what to do, but do not provide the power to do it."

In *Your Teen's Miraculous Brain*, you will get the tools you need to help your teen as you become better connected to the power to make lasting change happen. This is because as you acquire the tools you need to help your teen feel better, you'll begin to feel better yourself—as a person and a parent. The power that will help you make positive, productive, and proactive change happen in your life first will naturally flow into your teen's life, too.

The aim of this book is to join forces beyond commonly erected walls of neuroscience, mental health, and faith. Even if you think your teen's mental illness is untreatable, I've learned essentials to give him or her a fighting chance. But, like Moses picking up that snake by the tail, I will not promise you that these parenting essentials will be easy for you to pick up all at once with your teen.

Together, we can push through the fears of parenting through the teenage years into adulthood. Together, we can pick up what is good for you and your teen's brain and drop what is not. Together, we'll find many of those irritating, scary, and snaky problems you've been dealing with for all too long will turn into the very solutions you need to thrive as a parent of your tween, teen, or young adult. If you're willing to pick up what might right now seem to be your very own personal snake by the tail—those things that really bother and bewilder you about yourself, not just as a parent but as a person—I can promise you that you will find solutions to your tween's, teen's, or young adult's problems that you thought were impossible to solve.

Chapter 1

Soul Doctoring

Love cures people, both those who give it and those who receive it.
—**Karl Menninger**, American psychiatrist

The literal translation of the Latin word *psychiatry* means soul doctoring, but that doesn't sound much like the definition of that phrase. Today, psychiatry refers to a type of specialty doctor who works with mentally ill patients. *Soul doctoring* once implied a problem of body, mind, and spirit. Physicians once understood that mental health was often caused by more holistic physical ailments, such as imbalances of the humors or a spiritual cause, and was found in the center of thought, intelligence, and emotion—what we now know as the brain.

Parents, mental "health" experts are failing you if they are not taking into account your whole child—body, mind, and spirit. Psychiatry is failing you if, instead of getting to know what's really going on with your child, they send you away with a prescription medication and nothing else. Therapists, too, are failing you if their only prescription is for a behavioral change that fails to take into account the mind-body connection. What soul doctoring has come to is a false notion that either behavioral techniques or medication alone is the definitive answer for your child—when in fact it is only part of the solution.

Parents, modern medicine began failing you and your child when we began treating the brain, the body, and behavior separately. Although there are exceptions when medication is clearly needed for children, for the most part, medication to control behavior is not healthy for your child's developing brain and may cause lifelong changes to the next generation that none of us are truly prepared to grapple with. True, teens with a mental illness do need a psychiatrist and a therapist, but they also need parents who are not afraid to question what these professionals are prescribing for their teen and why.

"Intrusive" Parents

I admire how many parents of children who have a specialist like a psychiatrist or counselor insist on attending appointments with their teen rather than accept that the professional wants to minimize parental involvement. These parents resist those impersonal, more clinical counseling sessions and give their teen's health professional a wider perspective of what's going on at home that the teen may not be aware of, able to articulate or too ashamed to disclose. Although, at first, parent-child sessions can start out as gripe sessions about everything the teen or parent is doing wrong. Eventually, I find at the root of these complaints, most parents are concerned about their teen's basic health but the teen truly doesn't recognize how much their unhealthy lifestyle is seriously impacting their mood, thinking and future.

While the teen may fight for eating junk food, staying up late, and spending too much time on social media, the parents are seeing their teen heading for disaster in ways the teen cannot see. For instance, I'll hear parents say: "He's not sleeping." "She's a picky eater." "He's so anxious." "She can't stop moving." "He sits all day in front of his gaming computer." "She cries and laughs past midnight glued to that smart phone." And when it comes to teens, there has got to be something parents report about how disrespectful, dismissive, hurtful, and absent-minded their child has become since becoming a teenager. Then it's "My teen has such a bad attitude, I want to send him to military school!"

Wait a minute—that was my parents telling me when I was a rebellious adolescent that they were so done with me, they were going to send me off to boarding school! No matter what outward symptoms parents admit about their teen, the backstory to all these problems nearly always falls into exhaustion combined with parental guilt and frustration that nothing these parents have tried is helping.

Why Do Some Transition Age Youth (TAY) Fail to Launch?

Contributing factors from both parent and teen perspectives appear to cause young adults' failure to launch between ages 18 and 25 such as:

- Drug, alcohol, or other substance addictions.
- Gaming, social media, or other time-consuming behavioral addictions.
- Subtle or overt mental illness.
- Adverse childhood experiences.
- Negative influences by peer groups.

- Traumas experienced outside the family overwhelming their capacity to cope with adult stressors.
- Underdeveloped or immature brain functioning.

Why Do Mothers of TAYs Fail to Thrive?

In addition to those TAY factors just listed—which, incidentally, parents may also suffer from—some other factors that parents may contribute to their adult child's failure to launch may relate to:

- A parent-child misalliance is blocking the TAY from developing adulting skills.
- Parents have an emotional need to hold on to their TAY, and it's easier for the TAY to oblige by remaining dependent.
- Parents were so overprotective that the TAY didn't develop an ability to cope with adult challenges.
- Parents implied, "Do as I say, not as I do," by modeling poor lifestyle health themselves, leaving the TAY with no template for success of their own.
- Parents mistaking the TAY's failures as laziness, intentional defiance, or disrespect for the parents' efforts rather than immature brain functioning.
- Parents underestimating the power of their teen's miraculous developing brain by fearing that what they see now is what their teen will become.

Wherever blame is assigned, it's important to weigh the risks of these factors for either parent or child as they fall under four domains: mental illness, traumatic stress, substance or behavioral addictions, and brain development. To help you determine what's influencing your teen's behavior most and the risks these influences may hold, it's wise to enlist the help of a licensed mental health professional to determine if your teen requires professional intervention. No matter what the risk, the eight essentials outlined below will help put your teen on the path to greater independence.

No Matter What, These Eight Brain Essentials Will Help

Even if your teen sits in front of his computer, staring at a screen and drinking sodas all day, or she is on her smart phone constantly texting while unsociably morose with you and barely functioning otherwise, or she does not sleep well and has panic attacks because she is unable to cope with her overwhelming anxiety, or he is so depressed he rarely leaves the dark and dungeon-like existence he calls "*my* room"—there is a solution to all of these conundrums.

The answer is neuroplasticity, and brain health is the way to get there. Put simply, neuroplasticity is the brain's ability to change itself for better or worse. Changing the brain for the better is what this book is all about, not just for your teen but for you as well. As long as we're alive, neuroplasticity is always there for restoration, recovery, and repurposing the better brain your teen was meant to have. Your brain and your teen's brain have boundless power to solve nearly any problem that has been thrown at them because of their built-in ready and inextinguishable capacity for change, growth, and improvement.

I'll never forget an incident when my first two children were young. We were walking to a local bookstore. I was holding my two-year-old son's hand on a beautiful day in Berkeley, California, where I was attending graduate school. Suddenly my son tripped, fell, and broke his arm. I was so shocked looking at his bowed little radius bone. I couldn't believe it! I was holding his hand! All I could think was, *Mothers are not supposed to let that happen!* Frantic, I packed up the kids and got my son to the emergency room as quickly as I could. Everything seemed to be in slow motion. Finally, after checking in the emergency department and sitting down with my son in my lap, I'll never forget how mother's guilt overwhelmed me. *Could I have prevented this? Maybe if I carried him longer? How could he trip like that—come on, I was holding his hand?* Then like an overly dramatic animé character, words came with a burst of hot salty tears: "I'm such a bad mother!"

After the cast was put on and I was holding my son in my lap again, my son's attending doctor could see my anguish. He assured me that my son's break was the most common type of bone fracture he saw with children his age, and it tended to happen just the way my son's break did, tripping and falling forward while holding an adult's hand. Noticing my unrelieved apprehension, the second thing he said was, "You know, children's bones heal so fast, your son can practically grow a whole new arm as we speak."

While I'd like to say that the minute I arrived home from my son's emergency room visit I became a perfect parent, of course that isn't true. I have had times of good, mediocre, and sloppy parenting in the many years since that time. For example, much later, as my children were growing up, I started drinking a bit too much. I actually feel some defensiveness coming up right now from down deep inside of me because every time I think of being a "bad," "good," or "good enough mother," as British pediatrician and psychoanalyst D. W. Winnicott wrote in his famous book *Playing and Reality*, I think, *How can any of us parents ever really measure up?* Especially when life happened to our own good enough mothers and we didn't exactly get that Madonna-mother-of-God child-rearing ourselves.

As Winnicott points out so well, parenting is an interplay between perfectly attuning to the needs of our children as they mature and grow and allowing our children to gradually increase frustration for their natural progression toward independence. It sounds perfect in theory, but parenting is a much messier job than that. Sometimes it just hurts being a parent. After my son's broken arm and many mother-related victories and mishaps since, I completed my master's degree from the University of California, Berkeley; started practicing as a licensed psychotherapist; and completed a doctorate in integrated behavioral health from Arizona State University. Most of what I learned about helping others I learned from raising my own teens, my adult patients' recovery from their own childhoods, and translating all that into helping other parents—like you—raise their teens.

Optimal Brain Health

A truly evolutionary leap came for me when I rediscovered my passion for how the brain and psychotherapy combined to form a healing whole to better treat mental illnesses. This new insight started in a neuropathophysiology class in graduate school, in which I was first introduced to EEG neurofeedback and other ways of achieving brain regulation. Then I started learning about integrated behavioral and lifestyle health from my integrated health doctoral program. This combination of brain health, integrated medicine, and self-directed neurotherapy became my calling to help other parents like me solve many problems my patients face. These brain health–related problems include attention deficit/hyperactivity disorders (ADHD/ADD), anxiety, depression, oppositional-defiant disorder, posttraumatic stress disorder (PTSD), and developmental trauma. I found that the key to improving teen mood, thinking, and behavior problems is going directly to the source: their beautiful brains. When we get teens' brains healthy, they can change their world for the better.

In Max Lugavere's book *Genius Foods,* his physician editorial advisor, Paul Grewal, writes:

> Feeling the occasional bout of melancholy is a perfectly normal, and probably even healthy, aspect of the human condition. But if melancholia turns into negative self-talk, remember: don't judge your thought content or your mood unless you've been

> working out regularly. If you didn't walk your dog or let him out to play or run around every day, it would be considered animal abuse, and yet we seem to think it's okay to not move ourselves. Exercise should be the last thing to be abandoned when you're feeling busy or overwhelmed, not the first. When tested head-to-head against multiple antidepressants, three days a week of moderate exercise was determined to be equally effective as the pharmaceuticals, with the pleasant side effect of having zero side effects! Treat yourself at least as well as you do your puppy—you deserve it.

I add to Grewal's statement by reflecting that not only do you deserve the "treat" of lifestyle activation, but your grumpy, too often rude, seemingly lazy, stressed-out, and depressed teens and transition age youth do as well. In the next chapters, we'll help you treat yourself and your teen with even more brain healthy principles.

Integrated Medicine and Behavioral Health

In the last two decades, practitioners like Natasha Campbell-McBride, a leading pediatric neurologist and nutritional expert in the UK, have provided practical, tried-and-true advice on the gut-brain axis. In her book *Gut and Psychology Syndrome*, she details for other parents how she cured her own child with developmental and behavioral problems. By learning to improve her own son's gut health, she was able to pass this on to her patients. Campbell-McBride describes how the gut-brain axis has been overlooked because physicians have become too specialized:

> [D]ifferent medical specialties have been created, each concentrating on a particular bit of the human body…Mental problems are looked at from all sorts of angles: genetics,

> childhood experiences and psychological influences. The last thing that would be considered is looking at the patient's digestive system. Modern psychiatry just does not do that. And yet medical history has plenty of examples, where severe psychiatric conditions were cured by simply "cleaning out" the patient's gut.

Perhaps it's because the old idea was to literally clean out the gut rather than rebalance the good to bad probiotic ratio, but psychiatry tended to ignore the health of the second brain to mental and behavioral health. What practitioners like Campbell-McBride have found is that diversifying and balancing the psychobiotics in our gut is what has been working more consistently to help mental health patients, particularly younger ones, improve. Like these experts, I've learned that the one silver bullet is actually a combined effort of eight lifestyle health practices judiciously practiced together that make the difference in regulating mood, thinking, and behavior in teens.

Teens' and Parents' Brains Are "Wiring" Differently

Since teen and adult brains are constantly changing, Frances E. Jensen, neurologist and author of *The Teenage Brain,* writes that until recently, teen brains were not well understood. What neuroscience thought was factual about teen brains was proven fallacious. For instance, according to scientific thinking of the past, a child's intellectual capacity, including the structure and total lifetime potential of a child's intelligence, was considered complete by five years of age. Modern science believed that whatever we thought we measured as intelligence at age five was a child's total life potential. Essentially, we thought a child's IQ potential was essentially fixed at age five. Jensen states that the most damaging fallacy was that "Most people thought [the adolescent brain] was pretty much like an adult's [brain], only with fewer miles on it."

According to Jensen, "functioning, wiring, capacity are all different in adolescents." Granville Stanley Hall, the founder of the child study movement, said optimistically that adolescence was a time of "the birth of the imagination." He also said adolescence is a time of impulsive risk taking, moodiness, poor insight, and questionable judgment. Our misunderstanding of the teen brain has too often let teen behavior appear to parents as defiance. This misunderstanding of teen brains has likely caused many parents to discipline more harshly than necessary—or to take personally the affronts their teen's immature brains fling hurtfully at them.

The teen brain is particularly vulnerable to stressful life circumstances for several reasons. When stress hormones put teens on high alert, hormones that are already causing emotional dysregulation in teens intervene to worsen emotions and thinking. Under stressful conditions, sex hormones affect the teen's brain less predictably than in adult's brains. In teens, it's not just immature prefrontal lobes that cause emotional highs and lows; the same neurotransmitters and hormones that support adults' stress response returning to calm after an anxiety-provoking circumstance often work oppositely in teens.

For instance, while the stress hormone tetrahydropregnanolone (THP) is secreted to trigger the adult brain to return to a homeostatic calm, THP does not trigger a calmer state in teen brains. In the adult brain, when THP is released in response to stress, it helps reduce anxiety like a natural tranquilizer and takes full effect about 30 minutes after the stressful event has occurred. This helps adult nervous systems to return more quickly to a normal state of calm than adolescents. In adolescents, however, not only is THP ineffective in reducing anxiety, but it actually stirs up additional feelings of angst so that teens stay emotionally aroused longer than adults do even after an identical traumatic event occurs.

No longer operating under the defective assumption that a child's setpoint for learning, thinking, and emotional intelligence was fixed by age five, we now understand that the brain's potential has much greater potential throughout life. This means that right here, right now, however hopeless you feel your tween, teen, or young adult's problems may be, we now have ample evidence to show that you can promote positive neuroplasticity in your child that will improve their mood, thinking, and behavior for the better.

After considering these eight brain essentials, you'll start to notice that the antiquated and concretized ways of thinking about your child's brain remain steadfast among parents, teachers, counselors, and physicians. We now know that many of those frustrating, annoying, and worrisome teen behaviors are more a matter of immature brains with underdeveloped executive functioning dictating teen emotions and impulsivity. Once you've started on this mission to get your family more brain healthy, you'll see even what you thought was immaturity or even mental illness was largely due to a lack of basic brain health practices.

Chapter 2

Whose Life Is This Anyway?

Love the Lord with every passion of your heart, with all the energy of your being, and with every thought that is within you...And... love your friend in the same way you love yourself.

—Matthew 22:37, 39 TPT

As I strolled through my daughter's college bookstore recently, memories of my own college life flooded back to me. I loved everything about academia. Ginger lattes sipped over hours while poring over textbooks. The challenging clash of opinions. Exciting career opportunities. The possibilities of my life ahead. After two years of my dreaming big, my new husband graduated and started his first job, and my first child was born. I'll never regret having any of my children, but with the first came no doubt in my mind that for me, the carefree student life was gone.

It was time to start thinking practically. Self-indulgence had to become a thing of the past. No more classes just because they were interesting. Time to check off only coursework leading to a degree ASAP. My time was no longer my own. Confusion between what I wanted and what was best for my new baby constantly plagued me. Who would watch her while I was in class? How could I afford everything she needs on top of books and tuition? How can I work toward a career that would give me enough flexibility to earn a comfortable living but still have quality time with my children? All these questions added up to the reality that my own life could no longer be my most important pursuit. My dreams had suddenly been downsized to fit my reality—now I was a mom.

Shaking me out of those college daydreams, my daughter sheepishly peered out of the college dressing room. First I saw her beautiful tentative smile, and then her long golden tresses tucked haphazardly in a *toque blanche*. She was wearing a white student chef's coat with buttons bulging at the bust. Finally, fully exposed, she opened the door while the same time managing to grip her baggy *pied-de-poule* chef pants. As my daughter and I discussed how to resize this uniform ensemble required for her first culinary arts class, it hit me how somewhere along the way of raising four children, in many ways I had transferred some of my dreams of becoming this or that onto my children to pick up where I left off. Now, I was there in my daughter's college bookstore finding joy in her own excitement of preparing for her first culinary arts class. I was there loving my daughter in the same way I began learning to love myself long ago during my college years.

I hate to admit it, but I had not always been the kind of mother who took pure joy in the accomplishments of her children. For instance, my youngest daughter had said years before she enrolled in her first culinary arts class that she loved cooking and wanted to become a chef. I am ashamed to say that I discouraged her in high school. I thought the work was too hard—lifting heavy pots, demanding work conditions, and long, late hours cleaning up. I thought that male chefs still dominated the field and women were fighting for their right to a successful career. Not every woman could be a Julia Child, Clare Smyth, or Paula Deen. So, I discouraged her. She spent her first two years in college taking the science classes I advised her to take, but her passion wasn't there. She was losing interest in college. She was getting increasingly discouraged and anxious about her life ahead. I was baffled as to why she wasn't a carefree student like I was before I became a mom.

It's not always easy for me, drawing that line between mother and child. That line separates my life from theirs and marks the beginning of their own individuation. Their big decisions, not mine, about what's best for them. Their career path. Their choice to become Christians. Whether they would get married. If and when they'd start a family of their own. Most importantly, I had to become aware of what was best for my children when, too often, it was more about what was best for me. I needed to become conscious of those things hardest to see in myself—my own unfulfilled dreams. When it came to my children embarking on their young adult years, with each child, sooner or later, inevitably I had to answer the question: *Whose life is this anyway?*

Your Before and After Model for Change

When a religious scholar once asked Jesus what the greatest commandment was, contained within his answer, according to the Passion Translation of the Bible, were three words that describe much about the human brain: *passion*, *energy*, and *thought*. Similarly, child psychiatrist Daniel Siegel describes the brain as "the embodied and relational process that regulates the flow of energy and information." Passion, energy, and thought are what makes your brain and your teen's brain so miraculous, because both your brains are capable of neuroplasticity at any age. That means that by improving your brain health, you can ignite your passion, energy, and thought to change what you want for the better. Much of what we will explore is about just that, so that you are inspired to work through your teen's problems in ways you never thought possible. First, we'll start with your teen problem-solving model.

Assuming the model you are using now isn't working so well, we'll learn how you can disrupt what doesn't work and create a model that does work well for you and your teen. Life coach Brooke Castillo calls that model that isn't working so well right now your "Before Model." What are your worst fears, biggest worries, and major irritations about your teen or young adult?

____

____

____

____

Identify Your Before Teen Model

1. ***Now choose just one of those irritations or worries about your teen and describe only the facts of that one problem.***

   ____

   ____

   ____

   ____

2. ***Thinking about your past experience with this problem, what are your thoughts about your teen actually overcoming this problem?***

   ____

   ____

   ____

3. ***Given these thoughts, how are you feeling about this situation right now (resentment, fear, sadness)?***

____

____

____

____

____

4. ***What actions do you take or fail to take because of these thoughts and feelings?***

____

____

____

____

____

5. ***What has been the result of you acting and behaving this way with your teen in the past?***

____

____

____

____

____

Now let's do a kind of fact check on your before model for what Daniel G. Amen calls automatic negative thoughts (ANTs). Based on what I've heard from parents for much of a quarter century, parents' negative thoughts are the first most important problem for parents to correct. I've found that parents who don't correct their ANTs get themselves and their teen caught in a sticky web of misunderstanding leading to repetitive rebellious teen behavior or emotional withdrawal into depression or anxiety. The more parental ANTs, the less likely parents will see changes in their teens they hope for. So instead of allowing those ANTs to continue destroying the relationship you have with your teen, let's get them out of your way.

Ten Parental ANTs

ANT #1: Always or never thinking. You think something that happened before will always repeat itself. When you think in words or phrases like *never*, *always*, *no one*, *everyone*, *every time*, and *everything*, that always/never thinking is usually wrong. Example: *"He always takes advantage of me"* or *"She never listens to me, so why should I keep trying?"*

ANT #2: Focusing only on the negative. This occurs when you carry a long list of all the ways your teen is failing, while the real failure is more about you failing to recognize and affirm what your teen is doing right.

ANT #3: Fortune-telling. This is when you predict catastrophe based on what you see your teen doing or not doing now, as if he were doomed to continue this behavior the rest of his life. Example: "*If my teen can't remember to do these small things now, he's never going to succeed when I'm not here to remind him.*"

ANT #4: Mind reading. You know you need to stomp this one out when you believe you know what your teen is thinking about you but they haven't actually told you. Example: *"She thinks we're made of money"* or "*He hates me, and I'm just trying to help him.*"

ANT #5: Thinking with your feelings. This occurs when you fail to question the truth of your own feelings. Example: "*I feel like my teen does this just to hurt me.*"

ANT #6: Guilt beatings. happen when you start anything you say to yourself or to others with words like *should*, *must*, *ought to*, or *have to*. Example: *"I should have been a better mother so my teen would have behaved better" or "My daughter got in a car accident; I should have* *spent more time teaching her to drive and that would have prevented her accident."* If so, that's a good indication that you are taking too much responsibility for your teen's problem when sometimes it really is your teen who should be held responsible, not you.

ANT #7: Labeling. This occurs anytime you attach a label to yourself or your teen. When you commit a labeling ANT, you block your ability to be objective and fail to leave room for growth potential. Some examples of labeling your teen might be he or she is "lazy," "stupid," "a brat," "thoughtless," "totally irresponsible," etc.

ANT #8: Personalization. This ANT is most highly toxic for mothers. Personalization ANTs occur when innocuous events are taken as evidence that your teen dislikes, disrespects, disregards, or just does not love you. Example: "*If my son really loved me, he would call me more often.*"

ANT #9: Blame. Amen calls this the most poisonous ANT of all because blame causes you to become a victim of other people's behavior and, worse, causes you to believe nothing you do or have done will change your situation. Example: *"If only his father wasn't such a deadbeat parent, my teen wouldn't be so messed up."*

ANT #10: Vicarious. This ANT is insidious and very destructive to mothers and their children. If Melanie Klein, a well-known prodigy of Freud's psychoanalysis, had been into ANT therapy, she might say this ANT functions like her own idea of "projective identification." With this ANT, parents are attempting to satisfy their own unfulfilled needs or dreams through their children. Anything their teen wants to do that doesn't reflect their own unfulfilled dreams is criticized by them.

How Did These ANTs Get in My Head?

You might have identified an ANT or two in your before teen model, but how do you rid yourself of those pesky automatic thoughts? Just like unwanted ants in your home, you can try your best to fight them off, but until you get to the nest where the queen ant is, where those ANTs start, they'll just keep interfering in your relationship with your teen or young adult.

So try this: Find a quiet place alone. Get to wherever you do your best thinking. If you need to take a drive or a walk or just lock your bedroom door for some privacy, give yourself at least 20 minutes of undivided attention.

1. **Self-Care Check.** Like any good pilot or astronaut, do a preflight check to make sure you are clear for takeoff. Put your phone away, mute it but set an alarm so you don't check the time and get distracted with texts, calls, emails, or social media. Assess your own self-care:
   - ✓ Activity: Have you been active or sitting more than usual lately?
   - ✓ Nutrition: Have you eaten junk or fast food, overeaten, skipped a meal? Are you well hydrated?
   - ✓ Substance Use: Any unhealthy overdrinking or using recently?
   - ✓ Sleep: Did you wake up feeling rested this morning? Are you tired after a long day or long week of work?
   - ✓ Stress: Are major stresses unrelated to your teen looming large for you?
   - ✓ Love: Any relationship problems unrelated to your teen (including fear of loneliness) bothering you?

   Assess each of these. Even if you see major holes in your self-care checklist, its okay. Just tuck that information in the back of your mind and proceed with these as precautions. Sometimes ANTs crop up where they're not wanted because of poor self-care, so keep that in mind.

2. **Slow down and take a few deep, relaxing breaths.** Start thinking gentle thoughts about yourself and your teen. If you find your mind wandering to anything other than your teen-related ANT, write it down and assure yourself you will tackle that later. This exercise is about lingering with a few thoughts and the feeling that they hold longer than you might be accustomed to in your busy, hectic life. This moment is for you and you alone.
3. **Pick your ANT.** You may have a long list, but this exercise works best with red-hot fire ANTs, those that really hurt when they bite. So pick something that evokes strong feelings, such as fear, anger, grief, or loneliness, about your teen. The kind of feeling some might call a tipping point. Those ANTs that, if you let them keep biting at you, inevitably lead to hopelessness, helplessness, and feelings like you've reached your limit with your teen.
4. **Whose ANTs are they anyway?** When you think of your teen's problem, try asking yourself a few questions:
    - Do I need to feel like I'm a good mother? If so, is my teen's behavior evidence that I am a defective mother?
    - Do I need to know that all my sacrifices as a mother were worth it? If so, is my teen's appreciation or lack of appreciation evidence that I feel I sacrificed so much of my life for nothing?
    - Do I have a need to spare my teen the pain I felt when I was her age? If so, is my teen's struggle bringing up evidence that I have unresolved pain?
    - Do I have a need for my teen to have a better life than I did, and is his current problem causing me fear that his life might not be better than my own?

- Did I have a need to give my teen more of everything my parents did not give me, including time, money, and attention? If so, does it hurt that she doesn't appreciate that?
- Do I have a need to lean on my teen for emotional needs I'm not getting elsewhere? If so, do I find myself feeling rejected, jealous, or offended when my teen spends more time doing other things and not with me?
- Do I have unresolved hopes and dreams I unknowingly was hoping to see my teen accomplish for me? If so, do I find myself criticizing or minimizing my teen's hopes and dreams or lack of direction because they are not what I wanted her to do?

5. **Now, take a few more deep breaths and give yourself some appreciation.** After exploring these questions, what do you think you need from your teen that you actually might need to get for yourself? I know it's an odd question, but it's also a hugely important one, so mull it over a bit before you read on. Another way of asking this question is: Whose life is it anyway? Are some of your expectations for you teen better suited to be put on your bucket list instead of your teen's? Example: I notice I'm feeling sad about my teen. I take a deep breath and notice my thoughts. I might discover my ANT is something like this: *Every time I invite my teen to spend time together, she always has something better to do.* Where did this always/never ANT come from? Although spending time with my teen is important, could some of this come from needing more time with friends, spouse, or other family? Might a mom who gets really sad, angry, or resentful at their child for not spending enough time with her actually be feeling a loneliness that can't be filled by their teen?

6. **Okay, so maybe you've identified a few ANTs, but now what?** The way you get these ANTs out of your head is to disrupt your model with parental attitude training (PAT). For example, if your ANT comes from a vicarious or misplaced need like loneliness, the kind that's not your teen's responsibility to fulfill—great work! You've discovered that ANT; now it's time to PAT it. Now that you recognize that your sadness about your teen not spending more time with you is actually about you needing to reach out for friendship, ask yourself when the last time you called a friend, took a friend to lunch, or just got out socially at all was?

If you really want to improve your parent-teen relationship discovering your ANTS, finding out where they came from, and adjusting your attitude to meet your own unmet needs will go a long way to helping your teen. Just for the record, keep inviting your teen to spend time with you even if you have to corner him to do it, but make sure you also get your own needs for connecting with others met. Finally, be happy for your teen that he's connecting with others, even if it's often without you—which is developmentally healthy at his age.

Create Your After Teen Model

1. Now that you've cleared away a few ANTs, what do you want for your teen that he or she does not already have? Go ahead. Dream big.

____

____

____

____

____

2. What do you think and believe about your teen's future that makes it so awesome?

____

____

____

____

____

3. How does it feel to think about it this way?

____

____

____

____

____

4. How do you act and behave differently with your teen because of it?

____

____

____

____

____

5. What is the ultimate result you get when the dreams for your teen become reality, even if it's just in your imagination? How does it change the thoughts and feelings you're having about your teen now?

____

____

____

____

So far we've been working on how ANTs are thoughts, feelings, and unmet needs that negatively impact the mother-teen relationship and how to disrupt that before model. As we progress toward a better relationship with your teen, keep squashing those ANTs by adjusting your thinking and addressing any self-care needs you have. Now, it's time to add that third element that Jesus used, which also describes brain function: energy. Optimal brain energy comes from optimal brain health. The following chapter starts with a quiz to help you assess how confident you are at this point, in your brain healthy practices.

Chapter 3

Have We Launched Yet?

Stop imitating the ideals and opinions of the culture around you, but be inwardly transformed by the Holy Spirit through a total reformation of how you think.

—Romans 12:2 TPT

Brain Health Questionnaire

For each of these questions, please choose the number that corresponds to your confidence in creating a brain-healthy household (1=not at all; 10=totally confident).

1. How confident do you feel that every hour you sit you can take a few minutes to move and stretch, and then also get an overall minimum of 20 minutes per day (140 minutes or 25,000 steps per week) of **aerobic activity**?

Not at all 1 2 3 4 5 6 7 8 9 10 Totally confident

2. How confident do you feel that you can **eat nutritious food and stay hydrated** by drinking at least 64 ounces (eight 8-ounce glasses) of water throughout the day?

Not at all 1 2 3 4 5 6 7 8 9 10 Totally confident

3. How confident do you feel about improving your digestion with probiotics to **eliminate any gastrointestinal problems** you may have?

Not at all 1 2 3 4 5 6 7 8 9 10 Totally confident

4. How confident do you feel that you can get 7–8 hours of **restorative sleep** every night?

Not at all 1 2 3 4 5 6 7 8 9 10 Totally confident

5. How confident do you feel that you can control your use of alcohol, tobacco, cannabis, or other recreational substances and regulate other unhealthy **addictions** like overeating, gambling, overspending, or too much screen time?

Not at all 1 2 3 4 5 6 7 8 9 10 Totally confident

6. How confident do you feel that you can **manage stress** and still be productive by getting the things you want to do accomplished each day?

Not at all 1 2 3 4 5 6 7 8 9 10 Totally confident

7. How confident do you feel that you can talk with your teen about **sexuality and healthy intimacy** and recognize the signs that sex hormones might be hijacking your teen?
Not at all 1 2 3 4 5 6 7 8 9 10 Totally confident

8. How confident do you feel that you can connect with others, love and be loved, give and get help when you need it, and get your needs for feeling loved met?
Not at all 1 2 3 4 5 6 7 8 9 10 Totally confident

At this point, some of your ANTs may be coming out to bite you again with thoughts something like this: *If I can't get my teen to brush his teeth or take out the trash, what makes you think I can get him or her brain healthy?* Touché, good point, and glad you discovered that ANT. The best way to stomp out that dream-killing ANT is to make your home as brain healthy as possible. That is, your home becomes a place where everywhere your teen turns there are brain-healthy alternatives, including you! So, as Paul said to the Romans, "stop imitating the ideals and opinions of the culture around you" and start that "total reformation of how you think." As you create a more brain-healthy home, your teen can't help but become at least a little healthier, and it will form the foundation for fewer cranky moods and more reasonable thinking.

Essential #1: Brain Activation

In chapter 4, we'll discover why aerobic activity is the single most potent way of activating the brain at the metabolic, cellular, and molecular levels. These brain upgrades are made possible because aerobic activity stimulates the production of brain-derived neurotropic factor (BDNF). In addition to turning up neurogenesis (the process of nerve cell growth), BDNF also potentiates the genetic codes which together promote better cognition, memory, and longer life. True, all the reasons for becoming more active are infinite and nothing new; however, what you might not have considered is the latest science behind movement and aggregate brain function. Movement also moderates improvements in the brain by reducing damaging inflammation, increasing insulin sensitivity, expanding the size and capacity of the brain's memory center, and enhancing blood sugar control—much of this a direct result of exercise's accelerating effect on BDNF.

We'll explore ways to activate as you incorporate movement naturally as a time-saving part of daily life. Activating amounts to sitting less and standing, walking, and taking stretch breaks—just moving your body—more. Becoming brain activated is about becoming more activated in as many of the free spaces of time you have in between all the other still moments when you once sat. Even with the occasional sore feet and weary muscles activity can bring, when we live a more active lifestyle, we are not just transforming our brains; we are transforming the trajectory of our entire lives for the better. We'll learn the lesser-known benefits of physical activity to your teen's brain, including how activity translates into better moods and more positive and reasonable thinking—all amounting to your teen treating both you and the world around you with more care, congeniality, and respect.

Essential #2: Brain Fuel

In chapter 5, we'll discover the food-mood connection along with how to eliminate brain fog through super brain nutrition. Knowing the differences between whole food and food-like substances is knowing how food can be the worst and best medicine for brain healing, growth, and integration. It's not just your teen who gets grumpy when you make the wrong food choices; your relationship with your teen is strongly dictated by what you both eat. We'll explore brain nutrition, including adequate hydration, healthy fats, proteins, low chemically refined sugars, and carbohydrates—all infused with a wide variety of micronutrients. We'll also discuss the 2:1 ratio of omega-6 to omega-3 fatty acids. If your teen is a meat eater, we'll discuss how to increase your teen's sources of vitamins B_{12}, E, and creatine. If your teen is a vegetarian or vegan, we'll discuss what she may be missing in her diet, like creatine adenosine triphosphate (ATP), which is a brain booster during cognitively challenging moments. While refined sugars and most zero-calorie sweeteners are brain toxic, there are brain-healthy and sweetening alternatives.

Essential #3: Balanced Biome

If you've ever wondered why people say, "I don't know why, I just feel it in my gut" or "I just had that gut feeling," we'll have some answers for you in chapter 6. Every time we talk about intuition or having that gut feeling, it suggests the presence of a "gut" intelligence, better known as the second brain. We'll learn how much more accurate your gut feelings and intelligence can be if you have a balanced biome. Since 80 percent of our neurotransmitters are produced in our gut, a balanced biome means were producing more feel-good neurotransmitters, like serotonin, dopamine, and gamma-aminobutyric acid (GABA). Lacking any one of these feel-good neurotransmitters—and certainly having an insufficient quantity of all three—can produce nasty, mean, sad, and even suicidal feelings. We'll discover how your second brain's network of nerves has a bidirectional communication with your brain's central nervous system. Psychobiotics act like little communication coaches between your two brains via the gut-brain axis. Some psychobiotics produce and deliver neuroactive substances, such as GABA and serotonin. This means that certain psychobiotics provide natural antianxiety and antidepressant messages down neuropathways, like the vagus nerve, spinal cord, and neuroendocrine systems. Studies find people with irritable bowel syndrome (IBS) have a higher incidence of anxiety and depression than the general population without gastrointestinal problems, but more importantly, when their biomes were balanced their moods were balanced, too.

Essential #4: Sleep Restoration

In chapter 7, we'll discover why it's always two giant steps forward with good food and adequate exercise but three huge steps backward if your teen is getting not enough or too much sleep. If your teen is otherwise healthy, nightly insomnia or hypersomnia (sleeping too much day or night) will essentially erase all your hard work getting them brain healthy. Sleep is equally critical to brain health as exercise and nutrition. In fact, even if exercise and nutrition are in perfect balance, sleep is a capstone required to restore and preserve those brain gains.

If you think you've got this sleep thing handled with a pill, you may be tempted to snooze through chapter 7, but it's my hope that you'll expand your mind to the possibility that sleep is yet another bodily function best accomplished without pharmaceutical aids. Synthetic sleep holds hostage the good guys called glia, which are tiny players on your glymphatic system team—better known as your brain's cleanup crew. Chemically induced sleep disables the brain's natural detoxification process when, in the deepest stages of sleep, those tiny glia are working tirelessly, washing and organizing the brain. When you take a pill to chemically induce sleep, you are slowing down that glymphatic cleanup because medicated sleepers usually do not enter fully into the last two stages of sleep when glia get out their "mops" and do their best work. Later, we're going to debunk more obsolete myths about sleep and, in so doing, get you and your teen back to deeper, more blessed, and wondrous slumber.

Essential #5: Addiction Freedom

In chapter 8, we'll discover how becoming chemical-free means understanding that we don't need to ingest a mind-altering drug to accomplish something the brain and body can do much better with their own neurochemicals. The survival mechanisms of our brains can be trumped by unhealthy chemical and behavioral addictions. Both human and animal brains are prone to chemical dependency, but the truth is that you can remove the drug of choice and brains produce their own natural chemicals that keep your teen coming back for more gaming, promiscuous sex, and eating junk and fast foods (which can be potentially as damaging as drugs or alcohol). Conversely, the human brain is highly capable of becoming addicted to healthy behaviors that are actually better at producing feel-good neurochemicals; later, I'll show you how that's done. We'll also learn what is going on in your teen's head that makes them so emotionally and cognitively unstable and vulnerable to substance use. Teens' developing brains make them particularly vulnerable to addiction in all its many forms. In chapter 8, we'll also discover how addiction can work to your teen's benefit rather his/her detriment.

Essential #6: Productive Stress

In chapter 9, we discover why not all stress is bad. But stress affects teens much differently than adults, and teens usually don't have words adequate to express what the stressors are or how it all started in the first place, because the communications centers of their brains don't yet have all the five-bar cell towers we adults are supposed to have. Consequently, teens can't accurately identify what's going on in their heads. If the stressors started at a young age, this form of chronic stress is developmental trauma, and teens are much more negatively affected. When stress has been around constantly for too long, teens often appear too exhausted to expend the energy to keep above the gravitational pull caused by chronic stress, and these teens can slip into a helpless, hopeless depression. In chapter 7, we'll learn what's important for you and your teen to understand about stress, such as how to use stress more productively to promote brain health. While becoming stress-free is impossible, we can turn the tables on our stressors and finally make stress work more productively in our favor.

Essential #7: Sex Ed. Savvy

In chapter 10, we'll become more informed about those fabled raging hormones and how they affect teen mood, thinking, and behavior. Ignorance of these hormones casts aside teen's capacity to develop their executive reasoning at a critical window of opportunity for their brain development; when left to their own devices without parental guidance, it can be much to their detriment. As the author of *The Teen Brain,* pediatric neurologist Frances E. Jensen, says, "When it comes to teen hormones, the most important thing to remember is that the teenage brain is 'seeing' these hormones for the first time." During pubescence, the brain is experiencing a homeostatic crisis, a balance of power between hormones and other systems of the brain and body that are being greatly stretched and challenged.

We can thank estrogen, testosterone, and progesterone rising in our young women and men as the driving force behind these major changes. Sex hormones are why a teen can one minute be happy and seconds later double back to deep, dark despair or fits of rage. We'll also discover how you can communicate with your teen despite the hold sex hormones can have on their capacity to think rationally and relate respectfully to you. We'll also learn some helpful strategies to get your teen talking about love and friendships—even though often teens are at a loss for words to express how they feel without making you out as the one who caused each and every bad thing that ever happens to them.

Essential #8: Love Repurposed

In chapter 11, we'll discover why it doesn't always work to talk at teen brains like psychotherapists do or medicate teen brains like psychiatrists do. That's because teen brains respond better to their own internal sensory system, and self-directed neurotherapies invite the brain to engage with and change itself. Much like looking in a mirror and liking or disliking the image reflected back, neurotherapies clear away chatter about using the "right" words to describe feelings and thoughts that the mind may have about what the brain is actually experiencing and leaves the brain to speak for and correct itself.

As parents keep building on their teen's own neuroplasticity with seven core brain health essentials as a foundation, self-directed neurotherapies target specific areas needing mood regulation, calm and alert thinking, and purposeful behavior. We'll check out how three therapies, each in different ways, access the brain's potential to resolve problems on its own: eye movement desensitization and reprocessing, EEG biofeedback, and Internal Family Systems therapy.

To summarize what's ahead, the eight brain health essentials are:

1. **Brain activation** to birth the neurotransmitters and hormones that support the overall health of your teen.
2. **Brain Fuel** to provide hydration and macro- and micronutrients, and to help your teen's brain to build executive functioning and integrate overall brain function to its maximum potential.
3. **Balanced biome** to produce more feel good neurotransmitters and eliminate grumpy moods.
4. **Restorative sleep** to allow the brain's glymphatic system to recover and repair the teen brain.
5. **Addiction freedom** so the brain self-medicates naturally to regulate your teen's mood, thinking, and behavior with zero negative side effects.

6. **Productive stress** to strike the healthy balance between too little and too much stress and keep your teen in the Goldilocks zone—just right.
7. **Sex savvy** when parents give hugs, hold hands, provide loving eye contact to their teen and fear not those curious-about-sex talks.
8. **Love repurposed** when parents courageously explore their own internal conflicts and refocus on their teen's needs, teens can focus on principles, purpose, and meaning guiding their own lives.

The chapters ahead will download a lot of brain health information and present brain physiology principles that will inform parents how adult brains and tween, teen, and young adult brains function differently. This combination of brain health, brain functioning, and useable parenting practices will help your teen overcome mental illness, developmental trauma, addictions, mood, thinking, and behavior problems in ways that medication and talk therapy cannot achieve alone. In the chapters ahead, we'll also explore the importance of you, as your teen's parent, reparenting your own brain and your younger self.

Chapter 4

Brain Activation

"Miracle-Gro for the Brain"

If you are in a bad mood, go for a walk.
If you are still in a bad mood, go for another walk.
—**Hippocrates**, father of Western medicine

Liam's Brilliantly Busy Brain

Liam's mom is a super mom in every sense of the phrase and not at all an overdoing helicopter mom. She provides a great balance of taking all the right steps to see that Liam's higher-than-average intelligence is well preserved. The thing is, Liam has attention deficit/hyperactivity disorder (ADHD). His hyperactivity was out of control. As a young mother to her two boys, Liam and Chance (Liam's younger brother), Liam's mom has learned that not every childhood problem can be resolved with lectures, time-outs, and stronger boundaries. With a brilliant, busy young brain like her son's, Liam's mom discovered that she needed a team to help her son. Before starting him on low doses of ADHD medication, working with a child psychologist, and calling me to start brain health and EEG neurofeedback with him, Liam's mom had been constantly running interference with nearly everyone Liam came in contact with, even his preschool teachers—despite his overactivity, Liam's mom tried her best to keep him in the best school environments.

What confounded Liam's mom the most was that typical low dosages of ADHD medication were not improving Liam's behavior significantly enough to keep him in the classroom without problems. Liam had no trauma or other medical cause that could explain his unmanageable and unruly ADHD behavior. One thing was for sure: Liam's high IQ was not enough to help him control his hyperactivity and impulsiveness, but his mom did not want to overmedicate him.

Once Liam's team was on board with guiding them through better parenting, brain health, and EEG neurofeedback, to his mother's relief, his behavior became less misdirected and more purposeful. As an EEG neurofeedback provider, my part on the team was to help Liam speed up his slower prefrontal cortex activity and slow down his faster than average parietal lobes—just the way other brilliantly busy brains process information more efficiently.

Brain Health Essential #1: Brain Activation

Mental health clinicians have used movement as a symptomatic measure of depression, anxiety, and other mental illnesses for many years but had not yet clued in to the fact that lack of movement, or lack of it, was not just a symptom but a major part of the cure to that mental illness as well. For instance, when a patient is undergoing a mental health assessment, they are often noted to be moving slower than typical (a possible sign of depression) or faster than usual (a possible sign of anxiety). What we somewhat dense mental health clinicians have finally figured out is that activity level may be a cause as well as a curative factor to relieve anxiety, depression, and other mental illnesses. The benefits of being physically active are so miraculous to the brain that if you closed this book right now and got your teen out and active more, you would see miraculous improvements with this core brain health principle alone.

The reasons for getting active and exercising are infinite and hardly late-breaking news. We all know we'll live longer, reverse potentially life-threatening illnesses, and feel generally better when exercise is part of our daily routine. It's true that very large longitudinal studies on mortality show that less sedentary people live longer. Physically fit people also have fewer chronic illnesses during those longer lives. Less known are the benefits of physical activity to the brain and how improved brain health translates into the many ways your teen's brain will dictate more pleasant moods, rational thinking, and congenial behavior.

If your tween's or teen's problem is being too hyperactive or impulsive like Liam, brain activation will actually cause your child to become less impulsive and more purposeful in action with more activity. It sounds counterintuitive to suggest that an overly active kid should get more active. However, it is true that most activity actually calms down and regulates the brain. So, in addition to all the other awesome facts about exercise, brain activation improves cardiovascular stamina, builds muscle, and increases bone repair/density after illness or injuries. The more well-paced aerobic activity your teen gets throughout the day, the more likely he/she will sleep more soundly, retain more vital information, and better self-regulate his/her moods. As we will learn later, well-paced aerobic activity is an excellent addition to a stress management program; it helps reduce insulin resistance and inflammation caused by overproduction of cortisol in stressful situations.

Although aerobic activity is best for brain activation, all well-paced activity improves brain functioning, including interval training (aerobic), yoga (mindfulness, stretching, flexibility), weightlifting (anaerobic), or just sitting less and standing more. All these activities better regulate both the brain between your ears and your gut. Activity improves digestion in your second brain by promoting its ability to metabolize the water, fats, protein, amino acids, glucose, and trace micronutrients that optimize your teen's overall brain function. For your teen's mental health, the most important reason to move is that aerobic activity stimulates brain-derived neurotropic factor (BDNF), which naturally promotes new and healthier blood and brain cells.

Unlike other organs and muscles that take much longer to see the benefits of activity, the brain's miraculous neuroplasticity takes the experience of exercise and improves almost instantaneously in response to increased activity level. One of the most immediate responses is a change in mood, a more positive outlook, and seeing problems as less insurmountable. Because the brain responds almost immediately and positively to physical activity, it makes sense to start here to encourage you with quicker, more effective results.

The overall idea here can be summed up in four words: more activity, less sitting. Or, if you prefer a longer list, here's more: standing, walking, running, swimming, biking, cross-country skiing, hiking, dancing, horseback riding, competitive or noncompetitive sports—most anything that involves moving your body will do. Similarly, do less sitting, laying, lounging, snoozing (after a good night's sleep), loafing, chilling, and reclining because, except for sleep time, movement helps your brain regulate more efficiently.

Wait—I think I left out a major category of vegetating your brain: screen time. Screen time is sneaky, because while you trick your brain into thinking it's doing something, it's actually doing nothing much at all.

Brain-Derived Neurotropic Factor

BDNF has been called "Miracle-Gro for the brain." BDNF not only is necessary for creating new neuronal connections in the brain's memory centers (such as the hippocampus, located in the limbic system), but it is also responsible for developing and maintaining a healthy supply of the new neurotransmitters your teen needs to self-regulate. For instance, BDNF is necessary to produce serotonin (the neurotransmitter that helps us feel calm but alert), dopamine (which controls sensory-motor impulsivity), and oxytocin (the "love" neurotransmitter that helps us form attachments, trust, and bonding). BDNF is necessary to produce all the neurotransmitters we need to go about our daily lives.

Geeky guys in white coats have discovered that BDNF can perform remarkable feats outside your brain, too. Scientists have sprinkled a tad of BDNF on neurons in a petri dish, which caused those neurons to sprout new dendrites—the branch-like structures that reach out and make connections with other neurons, improving the brain's overall communication capacity. Sprouting new dendrites makes learning and remembering what we've learned possible. Without new dendrites sprouting in your child's brain, it's doubtful he/she could retain all the information needed to learn, mature, grow, and support the millionaire lifestyle you're expecting from him/her someday because you got their brain so healthy.

Another benefit of increased activity is that it stimulates the neurogenesis of neurotransmitters such as norepinephrine, which improves your child's productivity through increased memory and learning capacity. If your child develops a habit of breaking up study time with short exercise reprieves followed by a tall glass of water instead of a soda, the increased norepinephrine stimulated by exercise alone (no caffeine or sugar required here) will make him/her a much better student. Many ADHD trials have shown that children who participate in routine physical activity programs exhibit advanced cognitive and executive functioning in both math and reading. The more purposeful activity a child with ADHD engages in—that is, the more active and less sedentary they are throughout the day—the less likely they are to exhibit typical ADHD symptoms.

Boost BDNF before Boosting Medication

In her book, *Ten Facts You Should Know before Making the Decision to Medicate Your Child*, Mary Ammerman encourages parents to question psychiatric labels, because in most cases these labels are much like getting a diagnosis of fever—it often tells you what you already know about the symptoms without helping you understand the cause. Without diagnosing the cause correctly, the medication prescribed is given by trial and error to relieve some symptoms, but not all. Consequently, more medication is often prescribed, leading to med stacking. Children are being prescribed more medications at heavier dosages and at increasingly younger ages. The earlier medication is started, the more likely permanent changes will occur in young developing brains—changes that can set them up for failure in later life. By contrast, boosting BDNF is considered a critical component in relieving symptoms of major depression, schizophrenia, addiction, and other psychiatric and neurodevelopmental disorders. Even if you choose to medicate, "BDNF is critical to the mechanism of action of pharmacological therapies currently used to treat these diseases, such as antidepressants and antipsychotics," according to Anita Autry and Lisa Monteggia. The heathiest and most efficient way to boost BDNF is aerobic activity.

Childhood Anxiety and Activity

Anxious children benefit from how activity helps them regulate and balance their stress response hormones, such as cortisol, acetylcholine, and epinephrine (adrenaline)—all neurotransmitters that are activated by the amygdala and together form the hypothalamic-pituitary-adrenal (HPA) axis. The adrenal glands (located above the kidneys) cause cortisol secretion, which stimulates epinephrine production, resulting in your child's sudden flight, fight, or freeze responses (aka stress response). At this point, the hypothalamus sends the pituitary gland corticotropin-releasing hormone, which in turn causes the pituitary gland to secrete adrenocorticotropic hormone. If stress messages persist, the result is prolonged circulation of cortisol and a sustained stress reaction. This overactivating stress response is exhibited in your child's default stress behavior. More cortisol circulating in the blood often means more bad moods, objectionable behavior, and irrational thinking.

Trauma experts believe that because our survival mechanisms are still intact, movement is the next best way to lessen the impact of traumatic stress and the likelihood of creating childhood phobias. Moving the body tells the brain's HPA axis what to do with the sudden excessive adrenaline rush resulting from the stress response. When your child's coach sent him/her running around the playing field as a punishment for bad behavior, perhaps without knowing it, that coach was helping your child's brain produce BDNF, which in turn stimulated neurotransmitters like serotonin and dopamine to send calm and alert messages about how things will be just fine if he/she starts listening to the coach more next time.

Reduce Screen Time and Find the Time

We all know exercise is healthy for our bodies, but it's vitally important to mental health as well. Although I don't think I ever bothered to try earning the Presidential Physical Fitness Award as a kid, I do remember thinking it seemed like a cool idea watching other kids vie for it. What most of us didn't know way back then in the mid-1960s was how damaging too much sedentary time was, not only on our physical development but on our brain health as well. Perhaps this is why many behaviorally troubled children are annoyingly fidgety; their brains naturally want to move because it feels healthier than being confined to a schoolroom chair much of their day.

The thing we all seem to use as an excuse to not exercise is lack of time, so try reducing screen time and you will find the time you need to exercise. If you must have screen time, exercise while you're watching that screen. Passive screen time includes television, smart phone, tablets, computers, and watching movies—or just about anything that makes us an audience to life rather than a live actor on life's big, exciting stage. Often when we veg out in front of a screen, we're neglecting perfectly satisfying real time that's a part of our faith, family, friends, civic society, and peer support groups, to name a few. Sitting for 50 minutes with your psychotherapist can also be a good choice—especially if you have a lifestyle health therapist or coach who kicks you out to get your move on after sitting so long. Unless you plan on exercising more in front of a screen—maybe dusting off your old VHS tapes to start working out with Jane Fonda or *Sweatin' to the Oldies* with Richard Simmons—pry yourself away from every screen you are chained to.

Physical activation stimulates and regulates your teen's brain function. Getting more active is the single most important way to increase your teen's cooperation, motivation, and amiable moods. Activity is one of the three most important factors that will enable your teen to overcome nearly every setback he/she will experience for the rest of his/her life. Physical activation does all these miraculous things because it increases neurogenesis in the brain, the birthing process of new neurons, neurotransmission, and neuronal communication through dendritic growth and neural network connectivity.

Incorporate Ways of Being More Active with Your Teen

Lifestyle activation is not something you have to squeeze into a big chunk of your busy schedule as a major inconvenience, like going to the gym every night after work or joining a two-hour walking group at the break of dawn. Lifestyle activation is a way of slowing down incrementally, making more space in your life, stopping and smelling the roses more, walking your faithful Fido more, taking a flight of stairs when you could have taken the elevator, parking your car far out in left field when a space was available just outside the entry door, taking the scenic route powered by you, walking instead of sitting meditation, and standing when you had a perfectly good chair to sit in, even on the subway or at your desk.

Lifestyle activation is about looking for every opportunity to get out of your own head and on your own two feet. It's about not only moving more but engaging more with other living things, humans, animals, and plants. I have some patients who say, "Don't forget reptiles and insects." I have a teen patient who loves his pet water dragon (arboreal lizard) like most of us love our dogs, cats, and birds. Lifestyle activation is about strategically planning all the opportunities to move through life at your fullest pace of energy and grace. It's about becoming more animated and inconveniencing yourself a bit more by walking more and driving less for the greater good. It's about reversing at least some of the modern conveniences that got us all a bit fatter, lazier, and socially and emotionally more isolated and living more with less mechanical pampering.

Brain Launch: Activation

What unhealthy habit will you drop, and what brain-healthy tools will you pick up?

1. Children whose parents are active, especially if they exercise together, are more likely to report that exercise is enjoyable than children told by their parents to go exercise by themselves. If you and your teen are out of shape, it's best to start with 10–15 minutes of exercise per day, increasing by 10 percent per week as you reach your goal.
2. Tweens, teens, and transition age youth should work up to 60–90 minutes of activation daily; parents can benefit from as little as 150 minutes per week, but it's best to reach for a goal of 250 minutes per week.
3. By suggesting a walk-and-talk instead of a corner-and-lecture approach, parents can use activity to not only improve brain health but also redirect a parent-child argument that is going south.
4. Doctors inform us that when girls hit high school years, they tend to become less active, so a special nudge to get your teen girl's brain activated is important.
5. Introduce a variety of activities your teens might enjoy. Any aerobic activity jump-starts brains the best, but don't get stuck there; find movement your teen likes, including dancing, kickboxing, interval training (classic aerobic activity), yoga (mindfulness stretching/flexibility), weightlifting (classic anaerobic activity), competitive equestrian, biking, or kayaking.
6. Just beginning and ending your day with sitting less and standing and moving more will activate your brain positively.
7. If your child develops a habit of breaking up study time with short exercise reprieves followed by a tall glass of water instead

of a soda, the increased norepinephrine stimulated by exercise alone (no caffeine or sugar required here) will make him/her a much better student.

8. If no one in your family is moving, unplug them. Parents, you too. Overexposure to passive screen time is not only brain deactivating; it's fooling you and your brain into believing you have an active social life, while your family, including your teen, is missing out on the best version of you they have—the nonvirtual you.

No matter how you choose to get yourself and your teen into a more active life, remember that unlike formal exercise, activity—which is anything you do besides sitting and lying down—can be integrated into your daily routine. Doing almost anything in lieu of sitting and lying, even standing more, can accomplish this brain health goal each day.

Chapter 5
Super Fuel
Food as Medicine

Food is a powerful drug. You can use it to help mood and cognitive ability or you can unknowingly make things worse.
—**Daniel G. Amen**, psychiatrist

Lilly's Brain on Toxic Fuel

Lilly is a bright, beautiful, eleven-year-old diagnosed with attention deficit disorder (ADD) and autism spectrum disorder (ASD). Both conditions are likely caused by the developmental trauma of her prenatal and infant years. Even before birth, Lilly's start was rough; she was prenatally drug exposed, lived in a car most of her first 12 months of life, and essentially was severely sensory deprived. The good news came when Lilly was adopted. Lilly's adoptive parents were experienced parents who had raised four children, who each had successfully started lives of their own and were fully brain launched. Even though Lilly wants to load up on sugar and refined carbs, her mother now knows to steer her clear from all fast and junk food, including Lilly's favorite brain-unhealthy food: soda. Lilly gets exercise on the swim team in the summer and soccer in the spring. She sleeps well and has low stress in the home. Her family is close and loving. They practice a genuine faith, actively attending and serving in church.

Despite all that great parenting in a loving Christian home, Lilly's ADD and ASD symptoms were still overwhelmingly difficult for Lilly's very capable mother. Lilly's major problem was unexplainable brain storms with frequent flurries of venomous anger and physical violence. The only way Lilly's mother could deal with these episodes when Lilly was small was to hold Lilly in a tight basket hold, sometimes for hours, until her emotional storm was over.

I'll never forget when Lilly's mom and I first met. She said, "Please don't try to solve Lilly's problems with teaching me parenting skills. Not to sound like I'm bragging, but Lilly is my fifth child and I could write the book on good parenting—really. I love Lilly. Her family loves her. Her church loves her. Lilly loves us. We have formed a real attachment bond just as if I carried her myself and delivered her into the world. Her problem is her brain takes over, and we don't know why or how to stop these inconsolable brain storms she has had since we adopted her at nine months old. We know they are not seizures, and we know she's not just a spoiled brat, either." Working together, we found that the sources of Lilly's problem were a dysregulated brain and her reaction to the junk and fast food she had on occasion. What's remarkable about Lilly is that after getting her brain better fueled and 40 sessions of EEG neurofeedback one or two times per week, Lilly's brainstorms are becoming fewer and farther in between.

Brain Health Principle #2: Brain Fuel

The brain between your ears is estimated to comprise 60 to 80 percent water. With most of that water removed, what's left is roughly 60 percent fat, along with protein and trace micronutrients. Since water makes up most of the brain and body, getting and staying hydrated with good, clean H_2O is first and foremost for optimum brain health. According to a study published in *Frontiers in Human Neuroscience*, participants in the study who drank water before performing cognitive tasks reacted faster than people who did not drink water. In a must-subscribe blog, BrainMD, Daniel Amen's clinic reports that being dehydrated is the first thing that will slow thinking, elevate stress hormones, and impair judgment, and dehydration can damage your brain over time. The general rule for daily water consumption is drinking your total body weight in ounces divided by two. So, unless you are seriously overweight or underweight, you need to drink somewhere between seven and nine 8-ounce glasses of water daily.

Hydrating with plain water is far more helpful to your mood and thinking than adding sugar, caffeine, alcohol, or energy substances into what was once refreshing plain water. Sugar or not, sodas dehydrate delicate brain tissue because of their high sodium and carbonation content. Diet sodas are literally death to brain cells for many reasons, but the chief problem with diet sodas is that most sugar substitutes are diuretic and negate the hydration intake the brain would have received had you substituted water for that amount of diet soda. This includes any form of artificially sweetened beverages. Cane, beet, corn, and calorie-free syrups and powdered sweeteners have been linked to obesity, type 2 diabetes, hypertension, and cardiovascular disease in both children and adults.

Copious studies from multiple angles have not been successful in providing much impetus for drinking anything that tastes sweet—even calorie-free and reportedly "naturally derived" sugars. Read your juice labels carefully, too; often the worst possible sweetener, corn syrup, is added to juices. Whatever substance is making your drinks sweet, by any name—and there are a lot of them—is not good for hydrating your brain.

As for brains on alcohol, you can only hope that your teen is not sneaking alcohol from your kitchen cupboard when you're not looking or partying on the sly somewhere else. But depending on how much and how often you drink, alcohol is not making you such a brainiac, either. A neuroscience study of 1,839 participants with an average age of 61 found that, overall, the more alcohol participants drank over the years, the greater the shrinkage of their brains. Alcohol abstainers had higher brain volumes.

Polyunsaturated Fats and Fatty Acids

Healthy brains need healthy fats and fatty acids. In the past, it was thought that a ratio of 4:1 omega-6 to omega-3 fatty acids was optimum for brain and heart health; however, more recently, brain health experts are advocating for a more equal ratio of both or, at the least, striving for a 2:1 daily intake of omega-6 to omega-3 fatty acids. Good sources include extra virgin olive oil, fatty fish, krill oil, avocados, nuts, seeds, pasture-raised eggs, grass-fed beef, and leafy greens, to name a few. Two of the most important omega-3 fats are docosahexaenoic acid (DHA) and eicosatetraenoic acid (EPA). Of these two, DHA is most important as a structural component for building brain cells. EPA serves as an anti-inflammatory not only in the brain but in the whole body as well. Both omega-3 and omega-6 fatty acids are found in wild-caught fish, such as salmon, mackerel, sardines, krill, and certain forms of algae. They are also found to a lesser extent in grass-fed beef and pasture-raised organic eggs.

The great thing about omega-3 and omega-6 fatty acids is that some of the best sources are found in oils, seeds, and nuts and are easily disguisable in foods more palatable to your teen. If your child hates fish or refuses supplements, you can slip omega-rich foods in by finely grinding fresh seeds or nuts into their cereal, pudding, or mac and cheese. If your child has a texture issue with food, you can slip omega-rich oils on an organic cheese sandwich or make them a nut butter sandwich made with healthy fats like almond, peanut, or hazelnut butter. Kids hardly notice oil or ground seeds or nuts in healthier cookies you bake. If you don't have time to squeeze into your Betty Crocker apron, children often like superfood refrigerated bars without high-fructose corn syrup or refined sugar and made with nut flours instead of GMO wheat, corn, or oat flours.

Pasture-Raised Animal and Organic Plant Proteins

Studies have well established that diets lower in refined carbohydrates and higher in healthy fat and protein support better mood, thinking, and behavior in children. However, the controversy surrounding the nutritional value of protein sources you might choose in that better diet, such as soy and red meat, is still a matter of debate among doctors. The bad rap that beef gets related to cardiovascular and cerebrovascular health appears to be well founded, primarily in the majority of cases in which the beef industry raises cattle in unhealthy captivity (that sedentary lifestyle again) in those tight pens, the junk they feed cattle, and the copious prophylactic hormones, antibiotics, deworming medicines, and vaccinations they're routinely injected with. What does grass-fed beef (GFB) have to offer healthy brains? GFB provides essential minerals like iron and zinc, which are more easily metabolized in the human body than those coming from plant sources like iron in spinach or zinc in legumes. GFB is a major source of both omega-3 and omega-6 fatty acids and vitamins B_{12} and E, which are super brain micronutrients vital for mood regulation.

Creatine is prolific in GFB and acts as an energy regulator as adenosine triphosphate (ATP) rapidly cycles during brain draining cognitive challenges; creatine works in conjunction with ATP to improve the brain's memory and performance. When vegetarians and vegans, who tend to be low in creatine, are given creatine supplements, their cognitive performance improves. Other studies have shown giving people only 5 grams of creatine daily for six weeks supported better "working memory and processing speed and reduced mental fatigue in vegetarians," according to Max Lugavere. It's important to note that for omnivores, this creatine brain boost was not as significant, which may illustrate how the brain in some cases takes what it needs when it needs it most and doesn't "top off" its energy tank.

Like its animal protein counterpart, soy products went from queen of the day among plant-based proteins to villain in a decade. Unlike grass-fed beef, it appears that unless soy is ingested in its many fermented forms, it's a suboptimal brain food. Soy may weaken immune systems, cause potentially lethal food allergies, and counteract the positive effects of a well-balanced omega-6 to omega-3 blood ratio. These negative effects are due to soy products increasing the female hormone estrogen, which is also a problem with non-organic stockyard raised milk products. Organic and non-GMO soy is better but harder to find. However, soy is also high in potent enzyme inhibitors that may block the action of trypsin and other enzymes needed for protein digestion, which is thought to cancel out any potential benefit that soy may provide for those depending on soy as a prime source of protein. After considering all the downsides of soy, it is a relatively high source of plant-based protein compared to other legumes.

A Rainbow of Organic Fruits and Vegetables

In addition to providing essential micronutrients, including both soluble and insoluble fiber for the second brain, many vegetables and fruits have antioxidant properties that protect your brain against certain diseases. Increasing fruit and vegetable intake to five to 10 servings daily is shown to significantly reduce blood pressure, which may be a concern if your you or your child are on pharmaceuticals to manage mental health symptoms. The superfoods consistently voted most popular among doctors as their number one brain healthiest plant-based foods are again extra virgin olive oil, avocados, blueberries, and dark leafy greens.

Daniel and Tana Amen state that "eating from the rainbow" improves brain health by catching all the benefits of plant-based foods healthy for your brain. It gives us the best shot at obtaining brain-healthy nutrition because the colors of that rainbow of various fruits and veggies indicate the dominant micronutrients they contain inside their delicious natural packaging. In addition to eating from the rainbow, it's brain healthy to eat from a wide array of different animal and plant foods to promote biodiversity in our gut. According to neuroscientists, when we eat a wide range of organic fruits and vegetables, both fresh and fermented, and the other major food categories, we are also ingesting different pre- and probiotics contained within each different classification of food from vegetable to animal. The greater the diversity of the foods we eat, the healthier our microbiome community will be. As we vary the spectrum of food we eat, we're increasing the psychobiotics that in turn regulate better mood and thinking.

Whole Carbohydrates: Grains, Starch, and Sugars

What's the harm of a few guilty pleasures like soda, donuts, cakes, and cookies? As the highest concentration source of carbohydrates, sugar acts like a toxic jet fuel to the brain. It's added to every fast food and often hidden under the prepackaged savory, acidic, and salty flavors we love most. Hidden sugars are also disguised under 56 different names. I dare you to say just some of them without taking a breath. Now, breathe in...and here we go: fructose, sucrose, dextrose, lactose, glucose, maltodextrin, dextran, diastase, ethyl maltol, and galactose. Wait, don't take that breath yet! There are those easier to pronounce syrups too, such as cane, corn, artificial maple, carob, buttered, and brown rice syrups as well as non-cane-derived sugars like barley malt, honey, molasses, beets, and dates. If you prefer powdered sugars, they're found in raw, brown, confectioner's, caramelized, Barbados, and muscovado powders, to name a few.

Energy Drinks

According to a July 2018 report by the Institute on Health (NIH), boys and young men between 18 and 34 years of age consume energy drinks as their second most popular dietary supplement. Nearly 30 percent of children between ages 12 to 17 years of age consume energy drinks routinely. Despite the lure of the purported nutritional value, many of the "nutrients" these drinks contain are not proven safe for consumption.

In addition to large quantities of sugar typical in most soft drinks, what is also proven unhealthy for brain function is typically large quantities of caffeine contained in energy drinks which is equivalent to between four and five cups of coffee. The NIH indicates that in excess, caffeine has been associated with serious cardiovascular problems in youth such as palpitations, sleep disturbance, digestive problems, elevated blood pressure, and dehydration.

The natural additives typically found in energy drinks include: Taurine, guarana, ginseng, B vitamins, glucuronolactone, yohimbe, carnitine, and bitter orange. Guarana also contains caffeine therefore sneaking in even more than the average 500 mg of caffeine most energy drinks already contain. A 2011 study, found 42 percent of all energy-drink related emergency department visits involved combining these beverages with alcohol, marijuana, Ritalin, Adderall, and other pharmaceutical or illicit drugs. The NIH concludes that ingredients in energy drinks alone may increase teen's risk taking behaviors. If your teen complains of feeling anxious, panicky, nervous or is having difficulty sleeping, consider eliminating all energy drinks. It's likely a major hidden culprit in many of your teen's problems.

Clearly, the world has true love for sugar in all its Splenda—pun intended. We've only begun to count the ways here, and truthfully, your brain needs glucose to function. Since so many sugar sources are not brain healthy, what's left? Take that deep breath again—there are one or two sources of sweetener that are healthier for your brain, and the queen bees the world over are sitting on such a copious pile of it that these queens should be commanding super genius armies instead of buzzing insects. Organic raw honey is healthier for body and brain by providing a slower, more consistent metabolic release of glucose and disease-fighting probiotics. Slow release means consistent energy with fewer highs and lows. If you have a moody teen, refined and artificial sugars in their many forms should be the first thing you drop while you pick up and add organic, raw, local honey to your teen's diet.

Doctors have studied honey's brain-boosting potency and discovered numerous benefits compared to toxic refined sugars. Honey reduces cognitive decline with polyphenol properties that act to counter oxidative stress and restore memory deficits. Honey increases healthy estrogen production in women, which is needed for optimum learning and memory. When locally derived, honey's pollen content also helps relieve seasonal allergies. Most importantly to your second brain, honey contains probiotics that serve as superhero antibiotics supporting the whole body's immune system. For moody teens, I'll say it again: honey's slow and consistent glucose release promotes better mood regulation than refined sugars. Refined and artificial sugars do none of that; in fact, the corn sugars contained in so many "child-friendly" products are actually the most damaging to your teen's brain. If your teen is getting the standard American diet, which includes fast foods and carbonated beverages, he or she is getting a whole lot of brain-damaging corn syrup.

As an adult, your grains, starch, and sugar needs are much different from your teen's, especially if your teen is active in sports. Doctors agree that part of the muscle building needed for active teens includes the need to burn carbs. For teens to grow to their full physical potential, they require growth hormone to do its job in building and repairing the body. Growth hormone is deactivated by two things in combination: stress-driven cortisol and carb loading. Overconsumption of refined carbohydrates turns off growth hormone production. Carbs are necessary, and the best use of them is after a good workout because they help replenish muscle tissue. Just after a strength training, muscles become sponges pulling glucose out of the bloodstream, making carbs ingested after a workout less likely to be stored as fat. Incorporate those unrefined carbohydrates that teens need two or three times a week after a workout. There are many benefits to carb loading after a workout. The major benefit is turning up the fat metabolizing liver function with sweet potatoes, ripe bananas, berries, starchy veggies, and brown rice. This use of unrefined carbohydrates in their natural state also helps protect adult heart health by rerouting the storage of low-density lipoprotein (LDL) cholesterol instead of contributing to plaque buildup in your arteries.

Break-fasting

Eat a healthy breakfast, but also make sure you have *break-fasted* for your brain's sake. Brain experts have found that people who take a break from eating for a whole 12 hours between dinner and breakfast have healthier brains. The idea of fasting does not sound like fun, but it's easy to do and you probably do it more often than you think. Consider if you finish eating dinner at 6:00 p.m. and don't eat again until breakfast at 6:00 a.m.—this brain health tip is a snap. On the other hand, don't think you are even smarter by skipping breakfast; brain experts still keep finding that people who skip breakfast altogether are at higher risk for unhealthy body mass indexes, type 2 diabetes, and metabolic syndrome, despite actually reporting much lower total daily caloric intake.

The American Academy of Pediatrics recommends that children receive 24 to 30 grams of protein per day, depending on their age. It's also advisable that most of that protein be taken during early morning and afternoon meals and snacks. Teens in sports, highly active teens, and teens with ADHD/ADD should front-load their daily protein intake for breakfast and to sustain blood glucose levels, keeping them calm but alert until the lunch bell rings. In a 1998 study published in the *Archives of Pediatrics and Adolescent Medicine*, children who ate breakfast regularly had higher reading and math scores, lower anxiety and hyperactivity, better attention and focus, better school attendance, and fewer behavioral problems than students who did not eat breakfast. For children with ADHD, eating protein in the morning is especially important. In 1983, a study published in the *Journal of Psychiatric Research* tested three types of breakfasts—high carbohydrate, high protein, and no breakfast at all—among two groups, 39 children with ADHD and 44 children without ADHD. High-protein breakfasts produced the best effects, especially for children with ADHD. It is thought that children with ADHD often lack adequate levels of the amino acid tryptophan, which is an important precursor to neurotransmitters related to attention, learning, and self-control.

Fighting Hidden Neurotoxins

Until recently, doctors asserted that for otherwise healthy individuals, supplements were unnecessary. Since it was thought that most supplements were excreted through urine rather than used by the body, a common joke among physicians was that people who took vitamins weren't healthier; they just made more expensive urine. Today, most parents know that even when their children eat healthy to cover their other lifestyle bases, its best to take vitamin and mineral supplements. One major reason vitamin and mineral supplements are important to the mental health of your teen is many minerals and vitamins are needed by the body to detoxify all organs, including the brain. In addition heavy metals exposure, other neurotoxins, such as mold, Lyme disease, and toxic chemical exposure, can cause mental health symptoms.

Unseen, even naturally occurring environmental toxins such as heavy metals and other neurotoxins can overload the brain and insidiously impact teen mood, thinking, and behavior. William Walsh, an international expert in nutrition-based psychiatry, studied prisoners who were incarcerated for violent behavior and found that the most violent also had higher levels of heavy metals. Later, Walsh studied children in 2,400 families with at least one child with violent and disruptive behavior in the same family. The more violent children had much higher levels of heavy metals, such as lead, cadmium and mercury. Since both violent and nonviolent children were found in the same homes, it was discovered that the more violent children likely had a genetic component at work that prevented them from detoxifying like their less-violent siblings. He noted that the more-violent children lacked the nutrient blood levels needed to remove neurotoxins from their bodies.

Since those break through studies by Walsh, other research has linked mineral deficiencies including B complex, magnesium, methyl folate, vitamin D_3, zinc, and lithium to behavior disorders in children. Other studies show that insufficient or overabundance of trace minerals such as copper can become neurotoxic. Copper is needed in the production of neurotransmitters such as serotonin, yet copper in high quantities depletes zinc, which has been found to contribute to ADHD symptoms in children. Vitamins and minerals are important in fighting back against neurotoxins such as heavy metal poisoning. These hidden causes of mental illness are important areas of consideration for parents and health care providers when a teen seems treatment resistant to typical lifestyle, health, and psychiatric interventions.

Ask your teen's pediatrician about heavy metal, chemical, mold, and fungi testing along typical bloodwork that checks levels of minerals and vitamins as you are isolating what may be the cause of your teen's problem. Since it's impossible to eliminate all exposure to neurotoxins, its vitally important to make sure the body has what it needs to fight back by with chelation (detoxifying heavy metals) and other methods of neuro-detoxification. For more information on toxic mold see www.moldymovie.com.

Brain Launch: Fuel

What unhealthy habit will you drop, and what brain-healthy tools will you pick up?

1. How much water does your teen drink daily? Hydrate with water calculated at half of your total body weight in ounces daily (total body weight ÷ 2 = ounces of H_2O daily).
2. Is your teen getting omega-3 fatty acids? Balance a 2:1 ratio of omega-6 to omega-3 fatty acids from extra virgin olive oil, fish oil, fatty fish, avocados, nuts, seeds, pasture-raised eggs, grass-fed beef, and leafy green veggies.

3. If your teen is a meat eater, grass-fed beef provides a major source of both omega fatty acids (3 and 6), and red meat is an excellent source of vitamins B_{12} and E and creatine.
4. If your teen is a vegetarian or vegan, consider a creatine supplement, because it's scarce in plant-based proteins. Non-meat eaters who take a creatine supplement access this energy regulator that works with ATP to boost the brain during cognitive challenges.
5. Eat a wide array of whole foods from all food groups, both fresh and fermented.
6. When choosing fruits and vegetables, eat from the rainbow. Colors indicate the dominant micronutrients those delectable plant-based foods are providing your brain.
7. If you choose to stay with whole grains, avoid wheat and corn and do your carb loading after a workout to get maximum benefit. Don't expect carbs to boost your brain power except in the short run.
8. While refined sugars by all their names and artificial sweeteners as well are brain toxic in the quantities we get served in prepackaged, fast, and junk foods, get your sweet fix the queen bee way, with organic raw honey.
9. Fast routinely overnight for 12 hours and whole days on occasion. When you resume eating, always do only that one thing: eat—slowly and mindfully.
10. As much as possible, eat together with your family. Children who share family meals are healthier all around.

Chapter 6

Balanced Biome

The Gut-Brain Connection

If microbes are controlling the brain,
then microbes are controlling everything.
—**John F. Cryan**, neuroscientist

Elsa's Imbalanced Biome

Elsa has been the best at everything she does as long as her mother can remember. She plays leading roles in theater, sings like an angel, is a competitive athlete, and carries her team in academic pentathlons. Elsa is also kind and caring and wants to make the world a better place before she dies. The problem is that Elsa is only 14, but by day she had performance anxiety and irritable bowel syndrome (IBS), and by night she'd dream of failure and even have nightmares about death. Her anxiety got so bad that she started vomiting every day before school and before every kind of performance, from school plays to swim meets.

Except for her insomnia, Elsa and her mom thought that they were a brain-healthy family, especially because Elsa and her mom are vegans. Despite all that healthy food and exercise, something was causing Elsa to increasingly experience performance anxiety to a point that it was crippling her and she was afraid she would have to quit everything she loved: theater, band, choir, athletics, and academic competitions. After taking the brain health efficacy questionnaire, we found that one source of Elsa's problem could be an imbalanced biome. The IBS she suffered from wasn't just from anxiety. With a little investigating, Elsa remembered that she'd been on antibiotics for bronchitis, and it wasn't long after that her IBS got much worse. Antibiotics don't just target respiratory microbes. Another problem also may have been related to a lack of creatine, which is more common in vegans and vegetarians. After getting Elsa's biome balanced and adding probiotics, vitamins, minerals, and supplements like creatine recommended by her integrated medical physician, and Elsa and I using eye movement desensitization and reprocessing (EMDR) to overcome a performance trauma she experienced a few months back, Elsa's anxiety and insomnia are no longer preventing her from doing her best at all the activities she loves to do.

Brain Health Principle #3: Balanced Biome

Although your ear-to-ear brain is much more complicated, it's helpful to think of it as organized by three general levels: upstairs, or the cerebral hemispheres; downstairs, sometimes called the mammalian or the deep structure brain; and the basement—the lowest, most primitive region—called the brain stem. The top and outermost covering, called the cerebral cortex, particularly the frontal lobes, are where executive functioning occurs. When I refer to the deeper structures, I am primarily focused on the limbic system, our brain's emotional center. Below all the deep structures is the primitive part of our central nervous system, the brain stem, which is responsible for the autonomic nervous system (ANS) functions like breathing, heart rate, digestion, and other essential functions running without conscious thought.

A division of the ANS is the parasympathetic nervous system, or what has become known as the "second brain." The second brain is an extensive neuronal network that covers the gastrointestinal system like a sock, as unattractive as it sounds, from mouth to anus. The second brain has continuous two-way interactive communication with that big brain between your ears. Recent discoveries regarding the gut-brain axis have found that the second brain has a powerful effect on mood, thinking, and behavioral problems. Considering how most teens eat, it's important to carefully consider how your teen's mental health problem may be affected by how well the second brain is being fueled.

Improving your teen's gut health may in turn improve his or her mental health. Scientists have found gut-brain communication is so important that what you eat and how stressed you are literally cause your gut to dictate to your ear-to-ear brain what your mood and thinking will be. The second brain may be the single most important new discovery to improve mood, thinking, and behavior in your teen. And here's why—the gastrointestinal system contains psychobiotics.

The term *psychobiotic* was devised by neuroscientist John Cryan and psychiatrist Ted Dinan to explain the secret goings-on between the brain we all know about between our ears and the brain hugging our gut like a sock. Within those webbings of nerves around our "gut brain" that is communicating back and forth with brain #1, there lies a living culture. *Psychobiotics* refers to live organisms "that when ingested in adequate amounts, produce a health benefit for patients suffering from psychiatric illnesses." This secret rendezvous between our ear-to-ear brain and our gut brain affects us all, whether or not we are psychiatric patients. Also, there's another little detail: it's not just a few biotics—there is an entire community, more like a whole universe, of microbes like yeasts, single-cell protozoans, viruses, and bacteria, composed of bad guys (pathogens) and good guys (probiotics) making themselves at home in your gut and mine.

Part of making themselves at home in our gut involves a battle between bad guys shouting for takeout pizza and good guys sending hunger pang messages for good, healthy brain food. If we oblige the bad guys (pathogens) too often and habitually eat that highly refined carbo-loaded and glutenous pizza, they thank us with bad moods and foggy thinking after ever briefer happy feelings of hunger satiation. If we continue starving the good guys (probiotics), they will still fight on our behalf, though their numbers are dwindling, not unlike a Robin Hood of the microbiome. Probiotics will set siege to the bad guys even as they're starving for nutritious, brain-healthy food. When you eat unhealthy brain foods, good guys will still do their best at damage control on your behalf. That is, probiotics will continue fighting for us to keep the burgeoning troops of pathogens down. They continue to do their best to limit damage by communicating SOS distress signals through our second brain to brain #1.

The Food-Mood Connection

Ever wonder why people have been known to say, "I don't know why, I just feel it in my gut" or "I just had that gut feeling"? Every time we talk about intuition or having that gut feeling, it suggested the presence of a gut intelligence. So, when we talk about fueling the brain, we're actually feeding two brains. The gut brain sends messages up to the brain between your ears and also produces the majority of the feel-good neurotransmitters, like serotonin, dopamine, and gamma-aminobutyric acid (GABA). Lacking any one of these feel-good neurotransmitters—and certainly having an insufficient quantity of all three—can give your teen nasty, mean, sad, and even suicidal feelings.

It makes sense that gut feelings matter, since over 80 percent of our feel-good neurotransmitters are produced in our digestive system. A balanced biome means we're producing an adequate supply of serotonin, dopamine, and GABA as well as other neurotransmitters and hormones that help regulate mood, cognitive speed, acuity, and memory and buffer debilitating performance anxiety. Lacking any one of these natural feel-good neurochemicals can produce the kinds of anxiety and irrational thinking that was plaguing Elsa.

Your second brain's network of nerves has a bidirectional communication with your brain's central nervous system. Psychobiotics act like little communication coaches between your two brains via the gut-brain axis. Some psychobiotics produce and deliver neuroactive substances such as GABA and serotonin. Certain psychobiotics provide natural anti-anxiety and antidepressant messages down neuropathways, like the vagus nerve, spinal cord, and neuroendocrine systems.

When pathogens take over our gut, it's called dysbiosis, a term referring to microbial imbalances or maladaptation within our microbiome communities. When dysbiosis occurs, our biome sends messages telling our brain we feel sick, depressed, anxious, confused, nauseous, or all the above. Sometimes it's because we gave in and fed the bad guys, and as soon as we digest that whole unhealthy mess, we may get back to business, feeling okay. Other times we've picked up foreign invaders, cold or seasonal flu bugs, and our entire biome is under siege for as long as that illness is attacking. The thing is, some of the time it's all our fault we feel down, depressed, anxious, or unmotivated, because we literally fed those bad feelings.

When we suddenly feel sick, depressed, anxious, confused, foggy, or all that at once, we might want to take inventory of what we've been eating. If we've been feeding the bad guys their favorite foods—cake, pizza, cookies, hot dogs, sodas—and not what the good guys need—healthy brain food such as organic fruits and vegetables, fatty fish, pasture-raised eggs, white and dark meats—rather than sending us happiness and contentment messages, we're going to get grumbling and discontentment. Granted, sometimes we do feed the good guys good food, and pathogens or viruses from outside our biome invade giving us colds or flu. Whatever the cause, when pathogen colonies have the upper hand in our gut, spikes of angst will be sent to our central nervous system even before we feel physically sick and certainly as illness takes its course.

It's not just occasional sick messages that are sent up from our biome universe; it's also messages of how we should feel, think, and behave. Depression, anxiety, inattention, lack of focus, irritability, hyperactivity, temper tantrums, and rages can all start in our gut. This is not just conjecture on the part of a few mad scientists, either. There is a clear scientific connection between upset stomachs and your teen's irritability, anger, depression, and anxiety.

Feeding the Good Guys, Starving the Bad

The idea that we might get gut instructions about what we should think and feel from the "brain" surrounding our intestinal tract sounds like another fad—like in a minute, scientists will yell "Squirrel!" and we'll be led down yet another path to some other great new discovery to improve the mental health of our teens. I admit that the gut-brain axis sounds a bit farfetched, but it's an age-old concept dating back to the ancient Greeks. The father of medicine, Hippocrates, was right many centuries ago; we now just have better ways of proving he was right. There are multiple examples of a gut health–mental health connection.

Studies find that people with chronic ailments like irritable bowel syndrome (IBS) have a higher incidence of depression than the general population without gastrointestinal problems. Causality has been established; several studies of highly depressed patients with IBS who were otherwise resistant to traditional treatment regimens were dosed with fecal matter from nondepressed volunteers. The IBS patients with the added probiotics of nondepressed people were not only cured of IBS but also of their depression. And if you think this was a coincidence, psychobiotics like *L. bulgaricus* (found to help relieve depression) and *B. longum* (found to help relieve anxiety) have been compared with placebo in human populations, and the studies found that probiotics reduced depression and anxiety.

How Eating Right Improves Mental Health

So, how is knowledge of these two brains going to help you to help your teen? What's important here is that when we feed the probiotics and starve the pathogens, our teens will become more regular with digestion and elimination and also improve their mood, thinking, and behavior by positively impacting the three mood centers of our mammalian brain (the brain's limbic system).

These three mood centers are the: locus ceruleus (which produces norepinephrine), raphe nuclei (responsible for over 20 percent of overall serotonin production), and the ventral trigeminal area (which produces dopamine). Additionally, up to 80 percent of serotonin is actually produced in the gut, not in the brain. Probiotics will help your teen vastly improve their mood regulation through balancing overall brain processing with an adequate supply of serotonin.

The three feel-good neurotransmitters—norepinephrine, dopamine, and serotonin—send messages to your teen's ear-to-ear brain that they are capable, life is doable, and they can succeed. It's the gut-brain axis that is communicating to our whole system, body and mind, whether all is well or not so much, and psychobiotics do much of this two-way positive communication. If the idea of experimenting on your teen's microbiota concerns you, be assured that there are safe and healthy ways to introduce your teen to probiotics through either tested brands of probiotics in pill or liquid form or traditional fermented foods.

Fermented foods contain probiotics and many of the psychobiotics that can lower depression and anxiety while regulating mood and thinking. Since fermented foods—such as sauerkraut, kimchi, kefir, kombucha, miso, tempeh, and yogurt—have been eaten for centuries with positive benefits, they are time-tested and safe. Individual classes of probiotics have also been clinically tested as not only safe but targeted to treat specific problems. For instance, Probio'Stick was developed in France in 2008 and contains both *L. acidophilus* (R-52) and *B. longum* (R-175), which together have been shown to alleviate symptoms of those suffering from chronic stress. In a randomized, double-blind, placebo-controlled study, subjects taking Probio'Stick had increased cognition while depression and anxiety receded. There are many others, including VSL#3, Mutaflor, Align, Culturelle, Florastor, Yakult, and Activia yogurt. Many organic raw fruits and vegetables also carry probiotics that won't be lost if we learn ways of preserving them in our foods.

Probiotics That Improve Mental Health

We've peered into both brains a bit, and we've seen how probiotics can act as psychobiotics to regulate your teen's mood, thinking, and behavior. As a parent like you, with all this information, I would say, "So, how do I start with feeding this second brain to reduce my teen's mental health symptoms? How can I as a parent be sure that probiotics will improve my teen's mental health and not become just another wasted expense?" First, before we introduce any new probiotics to your teen's gut, it's best to start feeding the good ones he/she already has while starving the unhealthy biotics. There are many brain-healthy foods that feed the probiotics your teen already has in his/her microbiome.

The authors of *Psychobiotic Revolution*, Scott C. Anderson, John F. Cryan, and and Ted Dinan, write, "Just like your microbiota, your depression and anxiety are unique as you are—which makes it impossible to declare a singular fix for everyone. But most of us can follow a number of general rules and ways to unravel our special situation and take charge of our destiny." Although the potential number of probiotics are infinite, probiotics with the prefixes *bifido* and *lacto* are the psychobiotics that are known to "treat" mental health conditions common to what you and your teen might be suffering. Note that after the following psychobiotic names there is a strain designation, usually a number or combination of letters and numbers, that indicates this probiotic has been well tested and stabilized in a reliable form. That's one way of identifying the most reliable, safe, and tested probiotics for your teen.

BRAIN-HEALTHY CLINICALLY TESTED PSYCHOBIOTICS	
Bifidobacterium longum (R0175 & 1714) or *Bifidobacterium infantis* (35624)	Reduces anxiety and cortisol levels via its effects on hippocampal growth factors. Can reduce depression while increasing cognition by boosting antidepressant serotonin. Found along with *L. helveticus* in yogurt, kefir, and sauerkraut.

Bifidobacterium breve (1205)	Results similar to *B. longum*, but with slight differences as it has a stronger effect on anxiety than depression. Also prevents growth of *E. coli* and *Candida albicans*. Its strong antipathogen effect may explain how it relieves diarrhea, IBS, and allergies.
Bifidobacterium animalis (DN173010, BB-12, Bi-07) or *Bifidobacterium animalis lactis* (HN019, DR10)	Reduces IBS symptoms. Improves mood when used in combination with *L. bulgaricus*, *S. thermophilus*, and *L. lactis*.
Lactobacillus rhamnosus (IMC 501, JB-1, GG)	May reduce depression, anxiety, and OCD by increasing the neurotransmitter GABA and lowering corticosteroids, which reduces stress. Found in yogurt, fermented soy cheese, fermented sausage. *Note: Strong precautions for those with immune deficiency diseases.*
Lactobacillus plantarum (299v, PS128)	Enhances memory, even age-related memory loss. Reduces pain of IBS. Found in many fermented foods.
Lactobacillus casei (Shirota, DN-114001, Immunitas)	Reduces antibiotic-induced diarrhea and C. diff infections associated with anxiety. Studies of depression showed an improvement in mood after 10 days on yogurt containing *L. casei*.

Lactobacillus paracasei (IMC 502)	Lowers intestinal distress. Minimizes oxidative stress associated with intense physical distress. May reduce alcohol-related liver damage.
Streptococcus thermophilus	Synergistic with *L. delbruecki* in providing folic acid. Reduces hyper responses to negative emotion stimuli.
Saccharomyces boulardii	Only psychobiotic that is a yeast, not a bacterium. Recommended for anyone with compromised immunity or yeast sensitivity.

Our gut brain requires plenty of fluids, soluble and insoluble fiber, micronutrients found in vitamins, trace minerals, digestive enzymes, prebiotics, and the probiotics in your microbiome. To reduce overloading the brain's own detoxifying cleanup crew, the glymphatic system, these nutrient-dense foods should be fresh, organic, pasture raised, grass fed, including also some fermented and heated only when eliminating toxic microbes are necessary. In short, food should be consumed as whole as possible whenever practical without eliminating the naturally occurring probiotics that work so hard to keep you and both your brains—in your gut and between your ears—healthy.

Brain Launch: Balanced Biome

What unhealthy habit will you drop, and what brain-healthy tools will you pick up?

1. Eliminate junk food, fast food, sugar, white flour, and all other "food-like substances" to starve the unhealthy microbes in your gut.

2. Play "this for that" with your food. For instance, substitute brown for white rice, yogurt for ice cream, fruit for candy, fish for processed meat, and vegetables for chips wherever possible.
3. Add the all whole, unrefined, minimally processed, fresh, cage-free, pasture-raised, and organic foods you can to feed healthy microbes.
4. When taking antibiotics, also take a probiotic supplement that contains many different probiotics such as those starting with *bifido* and *lacto*, to replace the good guys your antibiotic kills off.
5. Promote biodiversity in your microbiome by eating a wide array of whole foods from plant and animal sources, both fresh and fermented, and consider a daily glass of water in the morning with a prebiotic supplement containing fiber.
6. Add fermented foods like unsweetened and unpasteurized yogurt, kefir, cheese, pickled fruits and vegetables, kombucha, kimchee, miso, coffee, chocolate, and more.
7. Drink less alcohol. Excess alcohol consumption can diminish the health, diversity, and balance of your microbiota.
8. Get plenty of omega-3 fatty acids in olive and nut oils, which reduce damaging inflammation in the brain between your ears as well as in your gut.
9. Choose fish with lower mercury levels, which includes those lower on the food chain like salmon, trout, tuna, sardines, and cod.
10. Avoid commercially produced emulsifiers such as carboxy methylcellulose (CMC) and polysorbate 80 (P80) and minimize your use of proton pump inhibitors (PPIs). If you're being treated for gastroesophageal reflux disease, IBS, or other gastrointestinal upsets, you're probably getting PPIs in your diet.

Chapter 7
Restorative Sleep
Your Brain's Night Time Clean-Up Crew

Less than 7 hours of sleep at night
causes lower overall brain function.
—**Daniel Amen**, psychiatrist

Brenden's Brain on Sleep Deprivation

When I first met Brenden, he had just turned 26 and was the most intelligent, verbal, self-aware, and clearly depressed young man I had met in two decades of practice. He was profoundly tired of living. Brenden was so deeply depressed that he said he even lacked the energy to end his own life. He was on high doses of medication for chronic pain, high blood pressure, and major depression, and worst of all, he rarely slept. When he did sleep, he woke up feeling groggy and overmedicated from his prescribed sleep aids.

A few sessions after getting to know each other, Brenden handed me a book about understanding the developmental trauma that adopted children experience. He asked me to read it because that's what he thought he needed help with the most. Even though Brenden's depression was more profound than many other patients I have worked with, surprisingly, Brenden's overall childhood trauma appeared low in comparison to other patients with major depression.

After I studied the book Brenden handed me, I came to understand Brenden's current depression. As an infant, he had virtually no attachment with either his biological mother or his adoptive mother. Instead, while awaiting being taken home by his new adoptive parents, he lay in the hospital, only fed and changed but with no consistent person to bond with. Brenden was left in the hospital without the human touch he so desperately needed in between feedings and diaper changes. When Brenden finally was discharged from the hospital and went home with his adoptive parents, he failed to thrive and lost so much weight as a newborn that he nearly died. I advised Brenden on the eight brain health essentials. I noted that he had completed a sleep study and was prescribed a sleep apnea device he was not using. What's remarkable about Brenden is that after he started practicing the core brain health essentials—activity, nutrition, and better sleep hygiene along with EEG neurofeedback—Brenden has stopped using pain medication, he has stopped smoking cannabis and drinking daily, and the weight of his depression extraordinarily lifted so that he can work through his early developmental trauma.

Brain Essential #3: Get Restorative Sleep

If you've ever seen younger children the morning after a slumber party, after a night of giggling, pillow fighting, spooky stories, and running amuck, with maybe a few hours' sleep, no scientist needs to tell you what happens next. Since teens' brains actually need more sleep than any other age category but typically get less, working out ways to improve your teen's sleep is imperative. Clearly, soundness of mind is dependent on soundness of sleep. In many ways, sleep is just as important as exercise and nutrition for building a healthy brain. In fact, even if exercise and nutrition are in perfect balance, if your sleep is off, there is no way to restore and maintain what your brain gained with either. Every lecture I have ever heard on sleep highlights three points: sleep is for rest, recovery, and repair of all bodily processes. A teen's developing brain requires sleep for cleaning and reorganizing their cognitive, emotional, and higher reasoning capacities. Most people do so much to disrupt their natural capacity to sleep that they think a good night's sleep is a hard science mystery resolved only with prescribed sleep medication. Here we're going to debunk that sleep myth.

When the book *The Physiological Problem of Sleep* hit the presses in 1913, I wonder if author and sleep scientist Henri Pieron envisioned it would take so many geeky guys in oversized white coats a whole century to figure out how important sleep is to our brains? In the 1920s, Nathaniel Kleitman identified sleep fundamentals like circadian rhythm, rapid eye movement (REM), and the miserable condition many of us experience from time to time known as sleep deprivation. Nearly 40 years after those awesome discoveries, the five stages of sleep were identified. In each stage—starting light, moving into deep and REM sleep stages, and then floating back up into light sleep again—there is a state of brain health restoration. Like practically everything else about our miraculous brains, sleep stages don't occur in a stepwise or orderly fashion. While we thought stages of sleep occurred something like descending a long set of stairs, getting some deep REM sleep down there somewhere at the bottom and ascending stages again until we woke up—we were wrong.

No, sleep stages, like many other brain processes, are just not that simple. Sleep stages are more complicated and not stepwise. Rather than a straight up and down staircase, sleep is more like Willy Wonka's futuristic elevator, the Wonkavator. As Willy Wonka describes, "The Wonkavator goes up, down, sideways, and every other which way you can think of going." Like that quirky but fun chocolate factory elevator, our sleep states go down and up, and sometimes sideways, four to six times throughout the night. Although we generally spend 90 minutes in each sleep cycle, that also varies, depending on age and certain conditions like sleep apnea.

If your teen is not a sound sleeper, he/she is not alone. Statistics range from 20 to 30 percent of children from infancy to adolescence struggling for better sleep. Teens are more prone to changes in circadian rhythm, delayed sleep onset, and inadequate total sleep time. Next to anxiety and depression, sleep is the number one complaint of adult patients that all physicians and integrated behavioral health practitioners like me hear. Furthermore, even among those anxious and depressed patients who didn't walk through my door originally to talk about how badly they'd been sleeping, just a little poking and prodding inevitably leads to some aspect of sleep gone awry. By the time patients tell their physician how badly they've been sleeping, usually they have been loading up on over-the-counter aids that haven't worked to keep them to sleep such as diphenhydramine HCl products (Benadryl, Tylenol PM, Advil PM). Physicians then prescribe hypnotics or the long list of off-label sleep aids like benzodiazepines, dopamine agonists, opiates, anticonvulsants, or an even longer list of other medications actually intended for psychiatric patients—*not* bad sleepers.

So nearly everyone who has a mental health complaint is also having a sleep problem. Some sleep too much, sleep too little, some wake up with night terrors, some have unknown nocturnal awakenings, and others have sleep apnea, but no matter what the sleep issue, we are all walking dead during the day without sleep. For nonintegrated physicians who still practice strictly Western medicine, that pill-for-every-ill mentality, trying to treat sleep with medication has become a hidden well of malpractice. This is because brains just can't get sleep from a plastic bottle full of little pharmaceutically compounded chemicals.

Do I hear you passionately protesting, "But that's not true! If it weren't for Dr. Xyzzy's prescription of Xanax, I would be doomed to a torturous existence, bloodshot eyes, eyelids fixed open, staring hopelessly like a zombie at my ceiling without a minute of sleep for all eternity!" But benzodiazepine drugs like Xanax are just the beginning of all the classes of medications that physicians prescribe in attempts to help patients sleep before trying the only brain-healthy way of peaceful slumber: good old unmedicated sleep hygiene. Permit me to list a few of them: anti-Parkinsonian drugs, non-benzodiazepine hypnotics, melatonin receptor stimulators, opiates, anticonvulsants, antinarcoleptics, and orexin receptor antagonists. Take a good look at that list again. Do you and your brain really want to get sleep that way? While it is true that people can take a nightly pill for a few weeks, maybe even a few months, and sleep like a baby, eventually our sneaky brains figure out what we're up to and, for the greater good, start waking us up again right out of that chemical coma and attempt to save us from the brain's eventual demise on drug-induced sleep.

Glymphatic System and Sleep Medication

If you remain convinced that you've got this sleep thing handled with a nightcap, cannabis, or over-the-counter or prescribed medication, please consider some critically important information regarding the health of your teen's brain. Your child's brain needs nonmedicated restorative sleep as much as he/she needs air, food, and hydration. No over-the-counter chemistry or prescription pill can give your child what nonmedicated restorative sleep can. Although we will go more in depth about medication soon, for now, I cannot emphasize enough that when you synthetically control what might look to you like your child sleeping, you are doing much more damage to him by medicating him into artificial sleep than suffering through a sleepless night. Start on the much better path of what doctors call sleep hygiene.

It is true that there are conditions in which it is necessary to help your child sleep with a sleep aid, but generally it's important to remember that human beings were made to sleep naturally, and by practicing good sleep hygiene, your child can get adequate sleep without a pharmaceutical. And it's highly important to your brain to try your best to sleep naturally. Pharmaceutically induced sleep in most cases circumvents, disrupts, or shortens deep sleep stages where your brain's cleanup crew, the glymphatic system, do their best work. When you chemically induce sleep, it's like you are giving the okay to a brain mafia to work your child's brain over after they have tied up the brain reclamation crew in the basement, making the good guys helpless to care for your child's brain by taking toxic waste away. Synthetic medications hold the good guys, the glymphatic system, hostage, while the brain's natural detoxification process is largely disabled.

Left unmedicated in the deepest stages of sleep, the glymphatic system—made up of tiny glia cells—works tirelessly washing and organizing your child's brain. These tiny glia work hard at night, pushing cleansing fluid through the brain's four fluid compartments: cerebrospinal, interstitial, intracellular, and blood vasculature of the brain. When you take a pill to chemically induce sleep, you are slowing down that glymphatic cleanup crew because medicated sleepers usually do not enter fully into the last two stages of sleep, and when they do, the time for brain cleansing is also shortened.

If you think your child's brain is sleep-resistant, take a minute right now yourself and notice how easily you can start to feel calm enough to begin feeling sleepy yourself. Lay your head back. Now, start taking breaths, slowly, deeply in through your nose and out through your mouth. Continue with a few deep cleansing breaths. Breathe more deeply and more slowly a few more times. Now scan your body and notice how you feel. You should notice you are relaxing at least a bit. You may notice you're feeling lighter, your muscles are unwinding, your joints are loosening, and your stomach tension is diminishing. If I placed electroencephalogram (EEG) sensors on your head right now as you lie back, relaxed, the little microphones listening to your brain activity would show me that your brain is now producing more alpha, theta, and delta waves than your brain was producing while sitting up intently reading this book.

Alpha, theta, and delta brain waves are long, slow electrical activity associated with calm states dominant during peaceful meditation, napping, and pre-stages of sleeping. In other words, your child's brain is designed to enter into calm peaceful states antecedent to sleep just by laying his/her head back and taking calm, relaxing breaths. My point is that the brain knows how to sleep without ingesting a pill; we just need to get that little plastic bottle of chemically induced intoxication off the top of your sleep solutions list.

Try Brain-Friendly Sleep Aids First

Natural sleep aids have been clinically proven to work over the short and long haul much better than pharmaceuticals. These are (1) L-tryptophan and theanine amino acids found in most foods like eggs, meat, poultry, fish, soy, seeds, and nuts; (2) valerian root, which increases GABA, and if used over time helps sleep onset and duration and quality of sleep; (3) chamomile, lavender, and lemon leaf teas, which, when served as a hot nut, seed, coconut, or dairy milk latte, also adds other micronutrients contributing to sleep; and (4) vitamins and minerals like calcium, magnesium, and potassium as well as calming B-complex vitamins. Using these along with all the other sleep hygiene tips as a comprehensive program supporting slumber will help you sleep more soundly. Although melatonin is widely believed to help with sleep, recent studies have debunked the myth that melatonin does anything more than aid initially falling into sleep. Melatonin does not help maintain sleep. What has been found to be more effective for longer sleep is melatonin's precursor L-tryptophan, which is rich in foods like turkey, almonds, and other plant and animal proteins.

Teens and Sleep

According to the National Sleep Foundation:

- During adolescence sleep patterns shift toward staying up and waking up later—which may explain why your teen likes to stay up later than you and is hard to wake up in time for school.
- Teens need 8–10 hours of sleep nightly to function best, but the average teen gets much less sleep.
- Teens typically have irregular sleep patterns across the week. For instance, on weekends and during school breaks they may catch up, but during the school week they return to sleep deprivation.
- Like adults, teens may suffer from treatable sleep disorders, such as narcolepsy, insomnia, restless leg syndrome, or sleep apnea.

How Does Screen Time Impact Sleep?

Let me count the many ways that screen time is annihilating your sleep time. Screen time is ravaging your sleep with blue light faking out your circadian rhythm, hypnotically suggesting what your dreams should contain, and causing you to depersonalize your opportunity for prayer, meditation, and setting your own intention during your brain's most precious time to commune with heavenly hosts.

First, blue light tricks stress hormones like cortisol to remain more active for longer while sleep onset hormones like melatonin remain inactivated for longer. Research has shown that blue light is particularly diabolical in disrupting sleep patterns because it suppresses melatonin production more than twice as long as other forms of light, thereby altering your natural light-induced circadian rhythm. One blue light exposure study found those whose screen time occurred between 9:00 and 11:00 p.m. had significantly shortened total sleep time, suppressed melatonin production, diminished sleep quality, and increased nighttime awakenings. Sounds like a bad night's sleep to me. Greater blue light exposure also prevents the body's natural incremental lowering of core body temperature, which corresponds with gradually progressing through various stages of sleep throughout the night. This study found that sleepers who are overexposed to blue light had core body temperatures that remained elevated to normal daytime levels and failed to reach deeper more restorative stages of sleep during the night. By contrast, the same study compared red light exposure and found none of these deleterious effects on sleep quality.

When patients hear those study findings they then say, "Okay, fine, I'll turn down the blue light exposure by dimming or filtering the blue light on the device or with a pair of blue light filtering eyeglasses." They think they've brilliantly solved that blue light problem regarding screen time, but then I bring up the second hindrance screen time presents before bedtime: the garbage-in/garbage-out issue. Honestly, fiction or nonfiction, news, reality shows, sitcoms, commercials, YouTube, Facebook, and even music often portray problems we all need to click off rather than on so that our sleep content can be more peaceful and productive.

This is what I call the sleep preview problem. Do you really want all that media garbage in your dreams? As a therapist who uses hypnogogic states to help people change negative cognitions and modify their constant self-debasing narrative, I am well acquainted with the concept of garbage in and garbage out. When you go to sleep with *CSI* on the TV, commercials repeating over and over with constant reminders that you in fact do not have everything money can buy yet, and you really should purchase this other thing now, subliminally you are setting yourself and your brain up for dissatisfaction because you now "need" things you had not yet thought you needed so badly before. And the news? That is a compilation of all the worst happenings on planet Earth in one large concentrated and highly discouraging toxic waste dump on your beautiful brain. If you're dumping the world's problems on your brain just before sleep, is it really any wonder your sleep is disturbed?

Finally, the most damaging aspect of screen time to college-bound teens' academic and psychosocial development is exposure to negative self-esteem content. A study of 4,508 students found that both screen time and content exposure had independent detrimental associations with school performance. Other studies found that adolescents with low self-esteem and high levels of rebelliousness or sensation seeking were those who also had increased media exposure. Again, these studies linked greater media exposure to higher-risk behaviors and worse school performance.

In addition to the daytime effects of sleep disturbance. children exposed to violent games, poor social values, and illegal or rebellious behavior just before they sleep are setting teens up for hypnotic suggestions. They are absorbing negative downloads into their subconscious sleep states that are like cement-filled shoes. By contrast, if sleep is used to help consolidate and store productive information going into sleep, they are given an edge to succeed academically and socially.

Sleep Is Foundational to Your Teen's Mental Health

The most fundamental reason why sleep is so important to mental health is directly correlated to the most difficult behavioral problems I see in the clinic with troubled tweens, teens, and young adults. Sleep deprivation tends to disengage the prefrontal cortex—the area of the brain we typically attribute to mature behavior and correspondingly well-thought-out and well-executed decision making. A major player in executive functioning, our prefrontal cortex usually puts emotional experiences into context so that we can respond appropriately. But when sleep deprived, the prefrontal cortex takes a back seat to primitive parts of the brain in the limbic system like the amygdala, our brain's "fire alarm" center, and we find ourselves more trigger-happy, sluggish, and yet hypervigilant. When we lack sleep, our primitive brain seems to take over, since the captain of the ship, our cortex, is essentially offline and our more primitive pirate's brain steers us into dangerous uncharted waters.

Smarter Sleeping

Although people tend not to recover all the sleep time they lost, they do typically recover the deep sleep they lost during longer than usual periods without sleep. Hence, your body's sleep system has some ability to make up for times when you don't get the amount of sleep you need. For the next several weeks, if you track your sleep, you'll notice that you occasionally had a relatively good night's sleep after one or several nights of poor sleep. Such a pattern suggests that your body's sleep system has an ability to make up for some of the sleep loss you experience over time. Doctors have dubbed this sleep banking, and in some occupations that require an inconsistent pattern of night to day sleep, sleep banking is an option.

If you are sleep deprived, you can take some comfort in more recent studies that indicate that sleep deprivation need not be permanently damaging. The important point to remember is that you do not need to worry a great deal about lost sleep, nor should you actively try to recover lost sleep. Needless worry and attempts to recover lost sleep will only worsen your sleep problem. Before attempting to change your child's sleep habits, it is important that you understand the effect sleep loss has on you as a parent. You will parent better after a good night's sleep.

Brain Launch: Restorative Sleep

What unhealthy habit will you drop, and what brain healthy tools will you pick up?

1. Maintain a predictable bedtime and awakening routine. Many people don't realize how their sleep times have drifted off their natural circadian rhythm.
2. Avoid depressants (alcohol), stimulants (caffeine), and psychoactive medications (like many antidepressants), and avoid large spicy meals four to six hours prior to sleep.
3. Exercise does deepen sleep, but not two hours before bedtime. Grabbing sun while exercising in the morning works to hit the reset button on your circadian rhythm.
4. Banish all unnecessary light and noise from your bedroom, including your devices—yes, even your digital alarm clock—and eliminate all blue-light-emitting screen time (TV, phone, computer) two hours prior to bedtime. Stop checking your devices if you wake up—that really messes with your circadian rhythm.

5. If noise is a must for you at night, start relying on white noise, which is low continuous or gently rhythmic tones like rolling waves or soft static. But turn off your TV, radio, and unless specifically designed for sleep. Turn off music, too. Put all media noise in another room. Remember, media noises were created to get your attention, not lull you to sleep.
6. If at all possible, keep the ambient temperature of your bedroom between 60 and 67 degrees.
7. Start rounding up and caging all those busy monkeys in your mind as you are preparing for your blissful night's sleep. Put all your worries, stresses, and business of the day to rest. One way to start stopping busy monkey mind is to become a list maker and give all those worries to the Creator. Every time a worry, an agitation, or overwhelming thoughts occur to you, before you plan to sleep, put your worry on your think-about-that-*mañana* list. Some find it helpful to set a positive intention about that worry and ask God to help them find solutions to that problem.
8. Although sleep banking does help your brain regain what it lost during a sleep deprived night or two, remember that your brain needs approximately seven to nine hours (teens need more, up to 10 hours) of restorative sleep each and every night.
9. The next time your well-meaning physician offers you a prescription sleep aid, do a Nancy Reagan: just say no. These include benzodiazepines, antihistamines, and brand name pharmaceuticals that are fallaciously promoted to give you a blissful night's sleep. Your physician means well but is thinking more about pleasing you than caring for your beautiful brain.
10. If you have had a sleep study and have a sleep apnea machine to ensure your brain gets uninterrupted oxygen all night and you are not using it, start. Remember that the benefits of using your continuous positive airway pressure (CPAP) or bi-level

positive airway pressure (BiPAP) machines will also help relieve your depression and anxiety. It's worth your time to consult with your sleep specialist about how to make using your CPAP or BiPAP doable for you. Adequate sleep reduces depression, improves brain health, decreases risk of dementia, and extends your life.

Chapter 8
Addiction Freedom
Your Best High Ever

When you have a persistent sense of heartbreak and gut wrench, the physical sensations become intolerable and we will do anything to make those feelings disappear. And that is really the origin of what happens in human pathology. People take drugs to make it disappear, and they cut themselves to make it disappear, and they starve themselves to make it disappear, and they have sex with anyone who comes along to make it disappear and once you have these horrible sensations in your body, you'll do anything to make it go away.
—**Bessel van der Kolk**, American psychiatrist

Ava's Brain on Psychotropic Medication

Ava is a remarkably beautiful 21-year-old diagnosed with no major physical problems and a long history of anxiety, panic attacks, and depression. When I first met Ava, she asked me to help her with her uncontrollable panic attacks and insomnia. She was told by her primary care physician that the Xanax she'd been prescribed for over three years to help her sleep and relieve her panic could be putting her at risk for developing dementia in later life, so she wanted to deal with her anxiety and be chemically free. Ava was the youngest of two girls and brought up Irish Catholic. She had a lot to contend with in her childhood, which reflected her mother's physical and verbal abuse of her, her father abandoning the family, and her mother's remarriage.

Ava recalls not having enough to eat or clothes to wear and too often feeling unsafe in her city neighborhood, where gangs were active. We discovered that prior to her physician prescribing a short-acting benzodiazepine for teenage anxiety, Ava had no history of panic attacks, just occasional normal anxiety. I asked Ava to consult with her psychiatrist about eliminating all antianxiety medication by titrating off with longer-acting nonnarcotic substitutes. Ava and I worked with the other brain health essentials. What's remarkable about Ava is that just by becoming brain healthy and getting off all inefficacious and unnecessary prescribed medication, she's sleeping more soundly, eating more healthfully, exercising more, and processing her trauma with Internal Family Systems therapy. Her fearful sad moods and panic attacks have nearly disappeared.

Brain Health Principle #4: Become Brain Health Addicted

The fourth principle to brain health is understanding addictions in a different, more helpful way than traditional views that we only become substance addicted. Humans do become chemically dependent, but did you know that the brain is just as capable of becoming behaviorally addicted as well? This can be good news if the habitual behavior you choose is healthy and bad news when the behavioral addiction is toxic. The brain becomes addicted and remains addicted even when the feel-good chemical or risky behavioral rush is over, and the consequences of the addiction become ever more dangerous. It's mind-boggling how dangerous addiction becomes when it grabs hold of someone you love—even worse when it grabs hold of you. Most of us have seen chemical dependency rip and tear someone else's life wide open. The real shocker is when, having seen the warning signs in someone else, you wake up to the fact that you are addicted. If you've suffered or are suffering from what drinking, drugging, overeating, raging, perfectionism, pornography, or the plethora of other ways addiction can hold you and your brain hostage, it's time to take an objective step back and view from it a brain's vantage point.

A brain's view of addiction demystifies, destigmatizes, and humanizes the whole addiction thing to a point—although that last part isn't entirely true. It's not just humans; monkeys get addicted to their kind of moonshine: fermented bananas. And primates aren't the only ones with addiction issues. Addictions are seen in wallabies on opium, elephants on the morula tree, reindeer on magic mushrooms, bighorn sheep on narcotic lichen, and jaguars on hallucinogenic yage vine. Scientists have even discovered the reason so many songbirds were dying in Vienna with cirrhotic livers was they were so addicted to fermented berries that they died flying into windows—and that's just addiction to a bunch of different kinds of moonshine! Our brains get addicted to toxic behaviors, too, whether beast or fowl, president or pimple-faced teen. Take a brain's view of addiction, and all its mayhem begins to make a lot more sense.

Thinking Like an Addicted Brain

Neurosciences continue to converge on findings in social sciences, and in the case of addiction, these doctors comparing notes have more or less solved the brain's addiction mystery. The funny thing is that what turned out to demystify the addicted brain is removing the drug of choice—alcohol, cannabis, street, and prescribed drugs. When doctors removed chemicals as a variable in their addiction studies and were left only with how the habit loop of the brain works, some astonishing discoveries got doctors jumping for geeky joy. Scientists looked at brain activity in nonchemically addicted subjects who had other habitual behaviors like gambling and found that the same pleasurable neurochemicals released by drugs and alcohol were—you guessed it—released in gamblers' brains with intermittent wins and losses. Apparently always winning would be too boring to the brain—not to mention getting you and your brain kicked out of the casino sooner than later. It was the thrill of suddenly winning along with whatever number of losing repetitions it required to hook a brain to create an addiction.

There is so much going on in the brains of tweens, teens, and young adults, and when on top of all that busy brain activity there is an addiction problem, objectivity becomes key. So, let's think like an addicted brain. All that is required to form an addiction is a stimulus strong enough to elicit the brain's pleasure-reward system. If you take all the unnecessary jibber-jabber—those confusing, hard-to-pronounce biochemical terms—out and just consider the brain's capacity to become and remain addicted, it's a somewhat simple and easy to understand process. When addiction hits you or someone you love, no matter how it's explained, addiction makes no sense at all. It's just overwhelming and sad. Mothers willing to give up their children, fathers going to prison for life, grandsons stealing from their nanas, daughters hooking for a fix—no clean and sober brain or doctor can ever really make sense of all that misery.

There is a solution. By backing off from moralizing, preaching, and shaming yourself and your child, you'll rebound from this heartache with a better start. In the case of addiction, rebounding as a parent involves you picking up that addiction snake by its slithery tail. As soon as you do, you'll see that your child's brain is addicted, and as long as his or her brain is craving that repetitive substance or behavior, that nasty snake will not be defeated. Lecturing, punishing, prodding, and beating your child up will only make addiction worse. When you see addiction from a brain's perspective, you'll come to see that addiction snake has become your shepherd's staff to lead you and your child safely home.

Teen Brains and Addictive Behaviors

Because of emotional and cognitive instabilities, teen brains are particularly vulnerable to addiction in all its many forms: drugs, alcohol, tobacco, food, excessive gaming, promiscuous sex, self-harm, and excessive risk taking—all behaviors that stimulate the pleasure centers of the brain. Understanding how any behavior can turn into a problematic addiction is helpful to parents of teens because the same process that straps your unsuspecting teen in for a dangerous ride on a methamphetamine run can become the same highly motivating brain process to get him or her craving healthy behaviors like working out at the gym or becoming a straight-A student. Both healthy and harmful habits are formed by the brain's own natural reward systems.

It's easier to get the fact that addiction is much less about becoming addicted to a specific ingestible or injectable drug of choice when we take the substance out of the equation altogether. When we do eliminate chemicals ingested from outside the body, we are still left with natural biological chemicals inside the body that work together to perpetuate addictive behavior. Any behavior can become an addiction; it's just that some behaviors place teens at greater risk than others. For instance, risk taking involving promiscuous sex, internet gaming, and cutting class can rapidly lead a successful teen to failure.

Higher-risk behaviors are more likely to ignite the addictive pleasure-reward centers of the teen brain. These addictions mimic the often tragic effects of chemical addiction that come from putting outside chemicals in our bodies to get high. This is because even chemicals we put into our bodies work on our own biological chemistry to inhibit or excite the brain synapses that activate pleasure centers. So, whether it's behavioral or chemical addictions, the one thing they have in common is they're both driving the brain's own pleasure-reward system.

Although a number of biochemicals are involved in getting hooked on a behavior or a substance, the chief enabling neurotransmitter is dopamine. Dopamine is both a neurotransmitter and a hormone that functions as an excitatory and inhibitory brain messenger. As a chemical messenger, dopamine helps motivate, drive, and focus the brain using its own reward circuitry, which reinforces goal-directed activity as well as the cravings perpetuating addiction. Consequently, when you're perplexed by how your teen keeps promising to complete missing assignments but instead he's completing another round of World of Warcraft, sneaking a peek at internet porn, or YouTube surfing again, what your teen is actually chasing is a bigger shot of dopamine rather than simply being too lazy or stubborn to make better choices with his time.

E-Cigarettes

While smoking traditional cigarettes has decreased dramatically among adolescents in recent years, the percentage of high schoolers who now vape using e-cigarettes nearly doubled in 2018. A typical e-cigarette dose, called a pod, contains as much nicotine as a pack of 20 regular cigarettes. In response, US Surgeon General Jerome Adams recently issued a strong advisory against the use of e-cigarettes among teens. Adams warns that "Nicotine is uniquely harmful to young and developing brains," and vaping "can cause learning, attention, and memory problems and can prime the brain for addiction in the future." Tobacco vapers may also be exposed to flavorants such as diacetyl (a chemical linked to serious lung disease) and heavy metals such as nickel, tin, and lead. The surgeon general concludes, "We must take aggressive steps to protect our children from these highly potent products.

Ensuring That Good Habits Stick

Life health coach Brooke Castillo demystifies how to drop bad habits while picking up healthier habits. It's a simple process of before and after. Take any habit you want change from five angles: (1) the facts of your current bad habit, (2) thoughts about your habit, (3) feelings keeping you stuck in that habit, (4) how you behave when you feel this way, and (5) the resulting action or reward you get for acting and behaving this way. That was your before. Then you repeat the same process, but instead of listing all the ways you are stuck in that bad habit, you project yourself into your picking up a healthier habit instead. As Brooke puts it, this puts your smart prefrontal cortex (the thinking part of your brain) in charge of the primitive part of your brain (which just wants to perpetuate that lazy, unhealthy habit) by connecting your limbic center (the place where your positive human emotions reside). When your prefrontal cortex joins with your limbic system and takes charge of your lazy reptilian brain, there is no limit to what you and your teen can accomplish. Brooke states that an important key is seeing failures as practice for success by staying curious about how the failure happened.

In other words, as we engage our prefrontal cortex (thoughts) and limbic system (feelings), we are exciting dopamine, a neurotransmitter that is key in addiction because it signals saliency, meaning that dopamine signals the prefrontal cortex to wake up and pay attention because a reward is coming. Brooke reminds us of Pavlov's dogs, those dogs that were trained to salivate at the sound of bell. That worked by pairing the sound of the bell with food. Eventually, the food was taken away but the dogs kept salivating. Why? Because of the repetition of pairing the bell and food, until bell meant food, which made the dogs salivate.

Like those dogs, once we are alert and focused on that vague notion that we felt good when repeating a certain behavior (like the bell paired with yummy food), saliency has begun to hook us into an addiction. While saliency is triggered more strongly when an addictive substance hits the bloodstream, conditioned cues like the curiosity that's triggered by smelling smoke as it rises from a group of teens with a look of euphoria from smoking cannabis or the hysterical drunken giggles coming from a college fraternity party pique attention, and the addiction begins. The neuropathways are much the same in the reward pathway when our saliency is triggered even when we're not ingesting a substance or repeating a dopamine rewarding behavior.

The key is saliency. It's like tucking in your dinner napkin before your dinner comes to the table because you are already salivating at the thought of satisfying hunger. Saliency brings up the sensations we associate with ingesting the substance or repeating the behavior. My point is that it does not take a drug to perpetuate addictive behaviors like risk taking, self-harming, gaming, or sexual promiscuity, but it does take saliency—that anticipatory, "Wow, I can't wait for that" feeling or "Mmm, that's going taste really good" to perpetuate both healthy habits and unhealthy addictions. The way we beat those bad habits is fighting back with the thinking and feeling parts of ourselves.

The problem with dopamine is that, like many chemicals ingested from outside your body, your teen can become more quickly tolerant to dopamine than other feel-good neurotransmitters like serotonin, because dopamine has a short half-life. This means that dopamine quickly causes positive feelings and then just as quickly drops out while other neurotransmitters become more dominant. It takes less than 10 minutes for a dopamine rush to wear off. Like drinking or drugging, in order to experience the same good feeling produced by dopamine, brains must chase the dopamine-stimulating behavior or chemical more and more to obtain the same good feeling it produced at first. If not, that depressing feeling of anhedonia, that dulling of the senses or missing the experience of pleasure derived from the chemical or behavior we previously found enjoyable, sets in.

Improve Mood and Thinking

Serotonin is a feel-good neurotransmitter that doesn't just give us a quick shot of feel good; it also gives a sense of calm and bliss that improves motivation for initiating and maintaining healthier behaviors. Scientists have found that when people are serotonin deprived, they are more likely to become aggressive and impulsive and make poor decisions, and they are less able to resist short-term gratification, less capable of future planning, and less interested in random acts of kindness. Bright light and vitamin D can attenuate some of these effects. Serotonin also aids sleep and is heavily involved in the kind of executive functioning that can stop an unhealthy behavior in its tracks. Increasing serotonin production to support better executive functioning in teens is worth your effort, and there are many ways to get more serotonin circulating in your teen's brain.

Antidepressants that act to increase the brain's use of serotonin categorized as SSRIs or serotonin-reuptake inhibitors cannot flip this trick for you, because no pill will give your teen more serotonin. What SSRIs actually do is cause the brain to keep serotonin active longer. In fact, if your teen's mental illness is not caused by a lack of serotonin and he's prescribed an antidepressant, there is an associated risk called serotonin syndrome, which can be very serious. Turning to a pill is not your teen's best brain health solution. There is no avoiding that your teen's two brains must work for more serotonin with lifestyle health, in particular, and super brain foods, exercise, and sleep.

Since serotonin is important to developing your teen's higher reasoning in his/her prefrontal lobes, it's important to know that 90 percent of serotonin is produced in the gut. Understanding how to boost serotonin as a preemptive measure against all manner of troubles your teen could fall into is another way to guard your teen against the pitfalls of transitioning safely into adulthood with an adequate supply in his/her brain.

One threat to serotonin and dopamine production is an endotoxin called lipopolysaccharide (LPS), a harmless gut companion that in typical quantities poses no threat to normal bowel function. The standard American diet, which is high in unhealthy fat and refined sugar, feeds LPS and suppresses the production of serotonin and dopamine. When gut balance is off, LPS colonies are the primary instigator of leaky gut and thereby become toxic to both serotonin and dopamine. The brain synthesizes serotonin from the essential amino acid tryptophan, which is also involved in the onset of sleep when it's synthesized into melatonin.

Improve Attention and Focus

Norepinephrine stimulates your teen's ability to focus, pay attention, and remember important details like turning in the homework he stayed up all the night before to finish. This is because norepinephrine's particular expertise is long-term potentiation, which involves storing and recalling those bad times so that we can avoid repeating that horrible feeling of failing again. If you want to model positive change for your teen, instead of picking up a glass of chardonnay, try picking up a new venture instead like learning a new skill or taking a new walking path while you talk with your teen.

Norepinephrine is also key in improving the brain's connectivity, but too much norepinephrine at inopportune times can be counterproductive by reinforcing fear conditioning. Norepinephrine draws us to potential danger but doesn't always help the brain distinguish between real or perceived threats. When we are in front of a screen, attempting to focus on something else, fact or fiction, whatever stress-inducing story that might be on, norepinephrine comes on line to redirect our attention and focus on whatever is in our environment that holds the most potential for danger. That's why avoiding the morning news before sending your teen off to school or getting yourself to work is a prime strategy to enhance attention, focus, and memory on remembering everything that you need to do before you both bound out the door, such as remembering where you left your car keys or your teen not forgetting to put his completed homework in his backpack.

As author of *Genius Foods* Max Lugavere puts it, "to mine norepinephrine for greater productivity…exercise." But why mine norepinephrine in the first place if it's only good for grabbing our attention to potentially dangerous stimuli? Norepinephrine is key in learning, comprehending, and retaining new information. Another way that Max recommends boosting norepinephrine to reset attention back to learning is giving our brains a jolt of extreme temperature change, either hot or cold.

Brain-Healthy Stimulants

As parents, we can keep healthy options open for stimulating those teen brain pleasure centers like combining activity and temperature, for instance, to get norepinephrine back on line for learning and creativity. It is stimulating your brain to take a leisurely walk on a perfect spring day, but don't hesitate to bundle up and get out and crunch down some freshly fallen snow on a chilly day. Take a hot yoga class. Work up a sweat in a sauna. Go singing and dancing on a chilly moonlit night, like Emma Stone and Ryan Gosling in *La La Land*. Those brain-healthy ways of taking your brain to the other side are much better than reenacting that admittedly awesome bar scene in *The Greatest Showman* in which Hugh Jackman and Zac Efron drunk-danced and sang. It may make exciting entertainment, but there are myriad much healthier, more exciting ways to reset your brain with happy neurotransmitters than drunk-dancing and singing. Get addicted to fun-induced healthy brain habits instead of drug-induced unhealthy addictions, and your brain can produce its own natural high, especially when your brain feels like it's stuck on depressive dumb.

What's also healthy about boosting norepinephrine, along with dopamine and serotonin, is that well-dosed and well-directed norepinephrine builds stress resilience by enhancing our ability to rebound from traumatic stress. If you want to direct neural chemicals like norepinephrine to work in your teen's favor, media fasting at least in the morning before school is a good part of your action plan. It will improve your teen's attention and focus. Rather than your family wasting time in separate rooms all watching their screens out of sheer boredom, a media fast makes them more likely to start taking walks and talking to each other more—which is good for teen brains, too. Become the change you'd like to see in your teen: consider using your brain's own addiction potential to work in favor of developing your own healthy habits and a much healthier, happier brain.

Brain Launch: Healthy Brain Addictions

What unhealthy habit will you drop, and what brain-healthy tools will you pick up?

1. Beat your own ineffective ways of coping (overdrinking, overeating, reliance on prescriptions, etc.) and instead do a brain hack that stimulates epinephrine with a challenging change in behavior, like take up dancing and hiking.

2. Stop being tethered to your devices at all times and instead replace unhealthy addictions with healthier ones in real time. You'll inspire your teen viscerally and profoundly to do the same.
3. Hold conversations that build your teen's executive functioning. First, gauge their stress level and don't raise it, because stress hormones like cortisol can stunt the neuroconnectivity from your teen's prefrontal cortex and lower brain centers that need to communicate with each other most to develop what we see in mature mood, thinking, and behavior.
4. Help your teen practice how he/she might avoid negative peer influences that might lead to unhealthy chemical or behavioral addictions or risky behavior. Maybe get your teen's views on big topics like war, peace, and whatever latest topic piques your teen's interest.
5. Make home life as safe as possible. Avoid creating an overstimulating environment. Avoid a chaotic home with too much criticism or a TV that spews a constant toxic background of media violence. In keeping home safe and calm, you're supporting positive norepinephrine production in your teen's brain.
6. Make sure home feels safe upon your teen's return from school so that her feel-good neurotransmitters like dopamine and serotonin remain adequate to begin rejuvenating her from the stressful day.
7. Commit to your own exercise, super brain nutrition, and sleep program to stimulate your own healthy supply of dopamine, serotonin, and norepinephrine. Getting brain healthy yourself is the next best thing to launching your teen's brain, so that you are also less inclined to default to unhealthy chemical or behavioral addictions.

8. Remember that there are many healthy ways of stimulating the brain's pleasure reward center. Long before chemical dependency has a chance to grab your teen's attention, make healthy brain habits an option available to serve as a buffer.
9. Become acquainted with the signs of substance use, and pay attention to your teen's behavior. Many parents have discovered that teens' unhealthy addictions came prior to a major failure in their teens.
10. If you believe you, your teen, or someone else close to you has a problem with drugs, alcohol, an eating disorder, or any other unhealthy addiction, get help. There are national referral lines, 12-step programs, and behavioral health programs to help.

Chapter 9
Productive Stress

Making Stress Your Ally

Real or imagined, when you perceive something as stressful, it creates the same response in the body

—Mark Hyman, MD

Sofia's Brain on Developmental Trauma

Without a doubt, my 22-year-old patient Sofia has been my greatest teacher in my 25 years of practice and over the last four years we've worked together. She is a remarkable overcomer and a brilliantly talented writer with a story of triumph to tell that is unprecedented in the quarter century that I have practiced. As I often tell her, Sofia is God's miracle child. Like most patients with severe developmental trauma, Sofia learned to cope by dissociating and breaking up into personality parts according to her age and the trauma, programming, physical pain, or deprivation she may have been forced to endure. Although she is diagnosed with dissociative identity disorder, you would not know it if you met her. Her dissociation is a healthy response to unhealthy childhood abuse and neglect. Her "parts" developed with specialized "jobs" to survive torture, rape, beatings, imprisonment, child pornography, and sex slavery—all, in Sofia's case, to help put food in their stomachs and clothes on their backs, and pay rent and utilities from what was left over from her mother's heroin addiction.

The unspeakably worst trauma for Sofia came with her mother's frequent participation in ritual abuse. I could write on, but I have learned to give hate, darkness, and depravity as little fame as possible so that the miracle of God's unfailing love shines brighter. Sofia's is a "But for the grace of God" story of survival. The first year working with Sofia, nothing I did therapeutically seemed to help. I tried it all, until along with the help of her church's deliverance team, Sofia and I began using Internal Family Systems therapy along with HeartSync prayer (www.heartysyncministries.org). Now, what's remarkable about Sofia is that she is brain healthier and claiming a life as a young, beautiful woman, which she never dreamed was possible as she looked out from the eyes of the tragically ritually abused child she once was.

Brain Health Principle #5: Productive Stress

Not all stress is bad. That's why it's both a number one killer and a number one bestseller. What most parents don't know about teens is that their brains are less equipped to cope with stress and more susceptible to its damaging effects. In the *Diagnostic and Statistical Manual of Mental*

Disorders, fifth edition (DSM 5), published by the American Psychiatric Association, there are three general categories of stress: acute, chronic, and posttraumatic. Acute stress is caused by an outside event with symptoms resolving within six months. Posttraumatic stress starts with a major life-threatening traumatic event triggering intrusive memories, flashbacks (day and/or night), and physically reexperiencing the fear and other symptoms, which are triggered seemingly without reason or warning. In addition to chronic stress, there is a fourth stress-related illness that is not yet in the DSM 5 and profoundly affects children and adults much more insidiously. This is the chronic stress caused by early childhood developmental trauma.

While acute stress hits hard and fast, its impact is the least disruptive to teens, as by definition, acute stress lasts no more than six months. Examples of acute stress might include your newly licensed 16-year-old excitedly pulling out of the DMV into oncoming traffic and totaling the old bombproof family car. No one was hurt; it was a lesson learned. Another example of acute stress might be your teen forgetting her lines in a school play. Although she felt devastated at first, she was able to recoup herself by performing better in the next play. Acute stress may knock the wind right out of your teen and throw them off-center for a few weeks, such that it's difficult to keep up with their daily routine, but they nevertheless bounce back to their normal routine in a relatively short time.

Chronic stress has a much different effect on teens, and they usually don't have words to express what the stressors are or how it all started in the first place. If the stressors started prior to age five, this form of chronic stress has come to be known as developmental trauma, and teens are much more negatively and insidiously affected. With both chronic stress and developmental trauma, teens' prefrontal lobes are less able to speak for the emotional centers of their brain than adults. When describing stress, teens can usually verbalize a few facts and it's obvious they are feeling something bad, but important details are lost in a fog of negative, viscerally deep feelings. From this side of the therapist's chair, watching teens trying to explain how chronic stress feels to them seems much like they're machete-slashing their way through an ever thickening jungle of tangled-up feelings, when the words coming out of their mouth don't make much sense. Their thoughts and emotions have been in such a tangle for so long that fighting to break free seems hopeless and immutable to relief. When stress has been around constantly for too long, teens often appear too exhausted to expend the energy on anxiety, and under the weight of chronic stress, teens slip into a helpless, hopeless depression.

By the time young adults are approaching 25, though it doesn't make the feeling any easier to bear, their frontal lobes are beginning to connect with language centers of their brain. This makes connecting with, describing, and resolving their feelings more possible. After 25, we can expect more young adults to have better insight about their thoughts, feelings, and behaviors; they are better equipped to explain how stress feels and how it started. For instance, one 24-year-old patient told me that the chronic stress she experienced due to developmental trauma felt like a black, thick, sticky, heavy blanket that suffocated the anxiety right out of her until nothing was left but sadness and depression.

When chronic stress exhausts us long enough, all thoughts, feelings, and motivation to change seem to slow to a stop, and even anxiety isn't worth the energy expenditure. Along with an absence of anxiety, so too a lack of motivation, activation, and the stimulus or drive to move forward fails with the inertia of depression. When depression hits as a result of the cumulative effect over time and intensity of chronic stress, stress is killing teens, not making them stronger.

Want to Become Stress-Free? Something Important to Consider First

Can a certain degree of stress be good for learning life lessons or activating attention and memory? We have biological examples to support both arguments that stress enhances learning and motivation but also is harmful in precisely the opposite ways. Stress can have both healthy and deleterious effects, depending on duration and dosage. For instance, your teen's gym coach might argue that the required sweat equity for earning that well-muscled body is stressing and breaking down less dense muscle mass so that muscle fibers regenerate into thicker, stronger, and more dense muscle mass (aka muscle building). But overhearing this conversation at your local gym, a cardiologist running on the treadmill beside you might say that your teen's coach has a point, but don't take his advice too far because chronically stressing your heart, which is also a muscle, creates all sorts of other problems more serious than what it's worth in getting a six pack.

If a third expert weighed in on this good vs. bad stress argument. Let's say you happen to chat with a neurologist doing laps in the gym pool. We might ask her if stress can build brains like it rebuilds better muscle. Before the neurologists answers this question, let's first agree on a working definition of stress and your brain. Stress is the brain's response to any demand. This means your brain responds to stress from everyday demands to big, life-stopping events. Stressors include everyday social pressures of work, school, family, or, for your teen, a hot new date prospect. All of that presents some level of demand on the brain to form a stress response.

Whatever the stimuli—pleasant, neutral, or tedious, depending on the day and the circumstance—it's all stressful. Whatever the stressor, our brains respond with compensatory strategies like dulling pain, improving memory recall, or sharpening cognitive performance to meet these demands. How damaging stress is to us depends on age of onset, duration, and intensity of the stressor. Chronic and developmental trauma, which is likely to have resulted from chaotic everyday difficulties in childhood, is more likely to lead to illness, because stress hormones released from the brain to meet stress demands are so continuous over time. Constant stress exposure accelerates oxidative damage in the brain and intestinal tract and throughout the body.

Teen Brains Are Wired Differently for Stress

In terms of major wiring differences, the teen brain has underdeveloped communication networks between the upper and lower as well as back and front brain quadrants in comparison to brains 25 years old and older. This means that teen's prefrontal lobes, which oversee better decision making, communicate poorly with central and deep structures of the brain, where more primal reactions and overwhelming feelings are produced. A case in point is how a teen's hypothalamic-pituitary-adrenal axis (HPA), located in the brain's limbic system, the center of emotional functioning, is unable to rely on better decision making like *stop*, *go*, or even *hold the phone, we've got to think about this before we respond.* Without the better connection between the HPA communicating with the PFC as is more common in adults, teens are at a great disadvantage in making decisions. They may hit Send before editing a text they may not have sent so quickly or at all, had their prefrontal cortex put a stop to it first.

The teen brain is different because the frontal lobes' wiring for communication is underdeveloped, and the amygdala does not run interference as quickly or accurately as it would in an adult's brain. In the underdeveloped teen brain, without accurate communication from their prefrontal cortex, the teen's amygdala (the fire alarm of the brain) is more likely to release stress-activating hormones that trigger the HPA process often when the stress demands do not call for such a big, dramatic response. The result is that the teen melodrama we see, which appears so manipulative and calculating, is in fact just a mishap of exposed wiring not yet connected to circuitry connecting the brain's more primitive protection and safety systems to the executive power of the teen's prefrontal cortex.

Without as much communication between emotion and reasoning centers as adults have, it takes teens much longer and much more cortisol circulating in the system before returning to a calmer state than what is called for given the circumstance. This means that teens overreact and stay that way longer than adults. Teens' overactivating and excitable amygdala is less capable of discriminating between what is and what is not truly dangerous or threatening. In late adolescence, particularly in girls, cortisol levels are slightly higher than in normal adults. This all amounts to what we outwardly see in teen behavior, betrays what's really occurring inwardly. Teens need adults to help regulate, calm, and sort through what the teen brain is learning to help them learn to regulate themselves.

Putting in a Good Word for Stress

A total absence of anxiety is not necessarily a good thing, especially when it comes to learning and memory. In the middle of all this emotional fallout, don't think you should lock your teen away in a stress-free bubble. There is a Goldilocks principle of stress all parents should understand when it comes to their teen's learning and academic performance. Two doctors, John Dodds and Robert Yerkes, experimented with arousal (anxiety) and performance (learning). They found that too much or too little stress is suboptimal for the best learning and performance. When levels of arousal became too high, learning and performance decreased. When levels of arousal were too low, learning and performance also decreased. But when a happy medium is struck between the two, learning and performance are optimal. This is because we need some anxiety to stimulate the glucocorticoids that are required to activate better cognitive performance.

Translated into parent speak: if you want your child to catch up or focus more intently on her homework and remember homework lessons for an upcoming test, it's best not to get into a verbal brawl with them about their academic performance while studying. If you want to help your teen with productive stress, it's probably best to turn the heat up just a tad, and then jump off the parenting soapbox before your teen experiences overwhelming stress. Let your teen deal with whatever problem you're trying to help her solve, but dose your encouraging or discouraging words carefully, and don't overwhelm them. If you want your teen to learn how to use stress productively, it's best you also learn productive stress thinking for yourself; otherwise, too much or too little parental effort may be counterproductive for you and your teen.

Stressed Out about Adulting?

Holly Swyers, a college professor, began noticing a new generation of her students, who feared that if they made one wrong move from the "right" path, the entire trajectory of their lives would be lost forever. They may have presented images of competence but privately felt helpless. According to psychotherapist Rachel Weinstein and former elementary school teacher Katie Brunelle, co-founders of the Adulting School, we have an epidemic among Millennials who are feeling isolated, overwhelmed, and paralyzed by a lack of their own basic life skills coupled with an abundance of choices at their disposal. Compounding their stress, parents can't figure out why their young adult children don't take action or, when they do, often fail to follow through. While their parents might argue it's their just deserts for doing too much for them, millennials counter that parents failed to come alongside them and teach them how adult life actually works. After graduating with high school and even college diplomas, parents are expecting their transitioning young adult children to somehow magically know how to make a living and fend for themselves without knowing basic adult life skills, such as all the many steps involved in landing that first job.

Problem-Solving with Productive Stress

Not all stress demands are negative. Stress in good measure is activating; it can boost our motivation to move forward or activate us just in time to avoid danger. If the brain is responding appropriately to meet the demand, it's not the stressor but our response to the stressor that causes an anxiety problem. Stress is negative when our brain sends a mismatched flight, fight, or freeze message and inappropriately responds to meet the demand. Like a lot of us, sometimes your teen needs space to mull things over. Ways a teen might tell you she needs space can be obvious, and we best not take it personally every time. For instance, if your teen yells to get you out of their room because they are too anxious to deal with too many missing assignments or if they fight to get away from you when you just wanted to give them a hug to comfort them when they're down, these are ways your teen is reacting to stress overload, so don't take it so personally and give them some space.

Most people suffer because of anticipatory anxiety or future thinking. Fear of the future only worsens anxiety as we freeze up in a stress ball and do nothing. When you take the action steps like the ones below, you can feel good that you made a well-thought-out plan of action. Action itself alleviates the worst part of stress, the helplessness, hopelessness, and worthlessness we experience when we are imagining the worst thing that might happen, which is usually much worse than what actually happens.

If you want to become the change you want to see in your teen, being proactive about practicing problem-solving skills as one of your stress management strategies will help you cope better and ultimately head off the impact stress has on your body before it takes its toll. Studies show that people who deal best with stress are those who have a wide spectrum of coping strategies, including many ways of finding fun and joy and dealing head-on and more effectively with stressful life events and problems. Doing what you can to problem solve, then accepting what you cannot do is key.

Brain Launch: Productive Stress

What unhealthy habit will you drop, and what brain-healthy tools will you pick up?

1. Recognize the difference between your teen's outward form—appearing lazy, stubborn, oppositional, or defiant—and what's likely really going on inside their brains. That's more accurately a short-circuiting of productivity caused by being stuck on stress.
2. Remind yourself that those emotional meltdowns are occurring between your teen's primitive and executive functioning centers. Don't add to the short-circuiting by being overly critical. When parents talk meanly, they contribute to their teen's sensory overload in flight, fight, or freeze mode.
3. You can reduce your teen's stress recovery time by not assuming the worst and zipping your lip until you have reason to know what's really going on with your teen.
4. By considering how your own stress load is affecting you and finding stress reduction practices that work for you, you can learn to use stress more productively.

5. Since stress is a part of life that we cannot eliminate, focusing on learning and practicing a healthy relationship to stress, rather than becoming stress-free, is key for you and your teen.
6. Learn productive stress communication; otherwise, you are likely to put too much or too little pressure on your teen—making your parenting efforts counterproductive.
7. For teens to swallow and then digest what you say, give them small chunks of information. Feed them in smaller bites—no more marathon lectures where you unload everything you want them to change in one long, one-way talk. Just stop it! Instead, say your chunk of information, then walk away and let your teen process whatever problem you're trying to help him/her solve.
8. Be proactive yourself about practicing your own stress management strategies. It will help you cope better and ultimately head off the impact stress has on your body before it takes its toll. Studies show that people who deal best with stress are those who have a wide spectrum of coping strategies, including many ways of finding fun and joy and dealing more effectively with stressful life events and problems.
9. To become a master of proactive stress, try this approach: identify the stressor, find that stress in your body, relax and reflect on possible solutions, choose what works best for you, go do it, and leave it alone.
10. Stop overthinking; instead, sleep on it. Much brilliant problem solving occurs in people who commit their problems to their wiser self, who awakens at night as monkey mind sleeps.

Chapter 10

Sex Ed Savvy

Learning the Curves on Teen Sexuality

When you say, "under no circumstances do I want you to do x" to an adolescent, x becomes the very thing they want to do... [instead,] treat them as people who have the capability to develop an inner compass.

—**Daniel J. Siegel**, psychiatrist

Elijah's Brain on Sex Hormones

When I met Elijah's mom, Eli just had become a legal adult with the developmental maturity of a much younger child. Eli's adoptive mom explained that along with his four younger siblings, Eli had a rough start in life. He and his siblings had a mother whose drug of choice was anything and everything she could get her hands on to snort, inject, or drink. When she needed to get high, she left Eli in charge for a few hours. During these times, Eli was the man of the house in charge of his younger siblings at the ripe old age of six, a position he'd had to learn how to do on his own around the time his youngest sister came along when he was only four. Then on one drug run, his mom was gone not just for a day or overnight but for weeks, leaving Eli and his siblings without adult supervision and care for so long that neighbors heard Eli and his baby siblings crying loudly and called Child Protective Services. Eli and his siblings were finally taken away to live in foster homes.

As too often happens, Eli and his siblings were split up and assigned to different homes. Unfortunately for Eli, he was assigned to a foster home where his foster father sexually abused him. Later Eli also admitted in therapy that he thinks his biological mom's boyfriends "hurt" him that way, too. Eli had also been prenatally drug exposed, making his way of thinking less mature socially and emotionally than his age mates in school but he did manage to academically keep up.

Eli's mom made sure he and his siblings always were active in church since they were young. Of her three adopted children, Eli was always the most helpful, mature, and obedient. He was always a great son, but hiding a serious problem. So, when she heard the earth-shattering news that Eli had been accused of molesting a family friend, a girl four years younger than him, Eli's adoptive mom had been hit with a catastrophe of epic proportions.

Eli's mom was shocked because she recalls many talks with her three adopted children about good touch, bad touch, and healthy sexual boundaries. She had them in therapy since their adoption and kept them all active in church. She even practiced brain health with them long before we met. What's remarkable about Eli is his adoptive mother's fierce love and dedication in partnering Eli through this transition age path he's on. She recognizes that though he appears physically mature, he lacks the cognitive and emotional intelligence to truly comprehend what he has done. Eli's adoptive mother is standing by him in ways his biological mother failed to from birth to six years old.

As parents, we can be remarkable like Eli's mother and help our teens negotiate the precarious journey through teen sexuality until they are more capable of making healthy sexual decisions independent of parental guidance. A great start in helping our teens is having a better understanding of the changes sexual hormones are having on their brains.

Brain Health Principle #6: Become Sex Savvy

About teen sex hormones, neurologist Frances E. Jensen writes, "When it comes to hormones, the most important thing to remember is that the teenage brain is 'seeing' these hormones for the first time." The brain is experiencing a homeostatic crisis, a balance of power between hormones and other systems of the brain and body that are being stretched and challenged. All the physical signs of puberty make more apparent the rapid changes in the brain that are also driven by sex hormones. A teen can one minute be happy and seconds later double back to deep, dark despair or fits of rage. Behavior, attitudes, and emotions change, as Dr. Jensen puts it, "all in the time it takes for her to close her bedroom door."

In both boys and girls, sex hormones are most active in the brain's emotional headquarters—the mammalian area of the brain called the limbic system. Testosterone in boys activates neuronal receptors primarily in the amygdala, a part of the limbic system. As the brain's fire alarm and switching station, the amygdala instantaneously takes over to reroute from passive, calm feelings of safety to an aroused hyperalert protective mode. Testosterone can increase as much as 30 times by the time a boy reaches full manhood, while estrogen and progesterone take the lead for girls. These hormones link to neurotransmitters throughout the entire limbic system of the brain, causing fluctuations in mood in rapid succession.

Sex hormones coerce teens into stimulus seeking, using brains that are not yet prepared for sound decision making, because in teens the prefrontal cortex—the brain's executive functioning—is not fully on line until sometime around age 25. In both genders, teens are not only more emotional but also more sensation seeking. This may explain why teens are not only more volatile emotionally but want to provocatively "grab" us for a ride with them on their emotional roller coaster. So, teens are more likely to start and maintain fights with parents. They are feeling intense emotional changes and tend to take risks to feel more of that intensity, like picking fights or speeding over the limit.

Although sex hormones are not a new discovery, today we have a better understanding of the ways sex hormones act on developing teen brains. We're discovering why the same amount of sex hormones is circulating in young adult brains as in teen brains, but unlike most adolescent brains, young adults are generally better able to control their emotional ups and downs. Why? It appears that the initial flooding of the brain's capacity to reorganize and regulate creates the roller-coaster problems with young adolescents. Stress can accentuate the intensity of biochemical reactivity, especially for younger adolescents. Greater stressors trigger stress hormones/neurotransmitters that intervene to stir up the already tumultuous regulation of sex hormones. Parents who become the change they want to see in their teens learn to reduce and manage their own parental stress so that their teens can better manage those raging hormones in a less overstimulating, calmer home environment.

Transition Age Sexual Identity

Although self-identity is constantly changing to a certain degree, social scientists find that between ages 26 and 30, most young adults have settled into a basic identity formation. Identity is composed of our morals, ethics, beliefs, self-other perceptions, goals, priorities, and body image—what's important and meaningful to us within our culture and community and among our peers. Sexual identity incorporates all that into a sense of self and body image. Our sexuality is a vital element of our identity formation, and in adolescence sexuality is newly resurrected from what Sigmund Freud called the latency period of childhood. Suddenly, sexual awareness bursts on the scene in pubescence. Without parental guidance, sexuality becomes a major rudder on the ship of individual human development that may set the course for a lifetime of pleasures or problems like teen pregnancies, STDs, and traumatizing effects caused by immature relationships.

Daniel J. Siegel explains the radical impact increased sex hormones have on both genders in the teen years. For teens, who are incorporating sexuality into their already amorphous but forming sense of identity, a major brainstorm has arrived, with sex hormones now online. Siegel explains further: "Now we develop a sexual self, a new identity filled with novel and powerful sensations. Some of these feelings may have been present earlier in life, but they were usually less intense, less persistent, and less available to conscious reflection during those years." Premature sexuality may force attachment injuries to resurface that began forming during the first years of life. Adolescents' adjustment and ability to incorporate a sexual self into their overall sense of self in relationship to others is very much founded on relationships from infancy.

If we experienced healthy, secure attachment bonds with adequate eye contact, a gaze that was joyfully sustained, we would in turn feel fully felt wrapped up in trustworthy, protective, and nurturing arms. If, instead, our attachment bonds were somehow opposing this gentle back-and-forth exchange between parent/adults and baby, we formed anxious-avoidant, anxious-ambivalent, or disorganized attachments, which later affect not only our adjustment as adolescents into adult sexual relationships but also how we lead as parents who want to be the change we want to see in our children. If we want our adolescents to experience a gentler and kinder transition to sexuality in adulthood, we need to become more self-aware about our own attachment experiences and the impact on our sexual identity.

Intimacy and Human Bonding

Sex sells, and we've sold out our teens' sexual development for the almighty buck. In many respects, by ignoring or shaming the sexual side of our humanness, religious institutions have harmed our teens by failing to meet the challenge of a new world in which objectifying sex is rampant. Left without healthy influences teaching about relational sex from parents at home or information at church, our children have nothing to fill the void but the lure of objectifying sex in the wide-open marketplace of mass media. It's remarkable how our children are so overexposed to media sexuality everywhere, and yet, for many families and among church communities, so underexposed to parental guidance about sexual behavior and relational intimacy in real time. Schools discuss the mechanics of sex and reproduction, but as a parent, apart from teaching diversity tolerance and mutual respect for teens from different sexual orientations, I don't want schools infusing my children with an education that usurps the natural development of their own sexual identity. Physical affection is best taught by modeling healthy relationships teens see at home.

When attachment patterns are unhealthy in a teen's primary family, many teens begin edging along an ever more daring ledge of sexual activity. Objectifying sex may start benignly with masturbation to pictures in mind of girls or guys and later actual pornography. When parental love and boundaries are not in place, teens may keep moving along a precarious continuum to cybersex, sexual encounters at school, and so forth, placing them at greater risk of sexual victimization before they have the experience to protect themselves. This unhealthy sexual continuum is beginning now at increasingly younger ages.

Objectifying sex may set teens up to coerce sex on peers or engage sexually with younger children. It's not just warning children about sex that makes the difference, as an adoptive mother of children who had been severely abused and neglected found out. Recall Eli's mother, who had done everything right, including keeping her children very active in a healthy community church, years of good counseling, all within a well-structured and loving home. Even with all those protective factors, children with developmental trauma and attachment injuries who had been sexually abused like her adoptive children remain at risk for unlawful and premature sexual activity.

Gender Identity

According the National Child Traumatic Stress Network, lesbian, gay, bisexual, transgender, and queer or questioning (LGBTQ) youth experience trauma at higher rates than their straight peers. These youth are bullied and harassed, suffer losses, and experience intimate partner violence and physical and sexual abuse at higher rates than straight youth. The social stigma of being different makes them more vulnerable to bias and rejection. Historically, poorly informed professionals fail to engage LGBTQ youth and provide ineffective treatments that, in some cases, perpetuate and compound the youth's traumatic experiences. If your teen is questioning his or her sexual orientation, your ignoring it, quoting Bible verses against it, shaming your teen for his or her experience, or failing in other ways to find support for your teen will not make it go away. You may disagree vehemently about your teen's interest in the LGBTQ community, but it's important to consider the trauma fallout that can occur in your teen, which could be at the root of your teen's mental illness. Failing to find compassion for your teen's sexual uncertainty at this vulnerable stage in life is likely to compound your teen's confusion and trauma and lead to gender dysphoria. It's important that you get qualified help for both you and your teen. Contact the National Child Traumatic Stress Network in Los Angeles (310-235-2633) or at Duke University (919-682-1552); website: www.nctsn.org.

Healthy Sexuality

Although we all are constantly exposed to objectifying sex in the media, in our homes most parents tend to ignore how overexposure to media sex impacts teen brains. Further, sexuality is a part of overall human health; however, many families don't discuss sex openly. I will not attempt to change anyone's mind about what constitutes acceptable or unacceptable teen sexuality, but I can say that, from adolescents to octogenarians, patients disclose to me problems they have in relation to their sexual selves routinely. So, unless you are comfortable with the idea that your teen may turn to professionals with views that may differ from your own, consider being the one to help your teen through these important questions about sex.

In my practice, I have patients wanting help to work through questioning whether they're bisexual or gay, transgender youth considering sex reassignment, and bisexuals who need to decide one way or the other because they are considering a long-term monogamous relationship with someone they love. I have seen many patients who have participated in long-term one-time or multiple extramarital affairs and patients who are averse to sex altogether. I work with those who have had torturous child sexual abuse and are left conflicted about sex because it triggers abuse memories. I have also worked with sex offenders, some of whom recognize the damage they've done and some who don't. Maybe because of a chronic illness or early childhood pain, most often I see couples in which one or both partners who have totally lost sexual interest but don't want to leave the marriage for a variety of reasons. I have seen several patients who know they are gay but decide to remain celibate and others in partnership who want to be accepted in the religious affiliation of their choice or have experienced rejection by the religion of their childhood and family.

For your adolescent's sake, if you are of a faith that considers masturbation a sin, you may want to question that as a parent. From a mental health standpoint, self-sex is common and only creates a problem if masturbatory behavior is coupled with pornography and/or drug use and/or becomes a time-consuming addiction. Whether it's internet gaming, talking, texting, or masturbation, any behavior that distracts from your teen's functioning in school, home, church, or healthy social life is a problem. Despite more openness about sex, adolescents still carry shame and guilt about their private sexual behaviors, so it's important to decide where you stand on the issues, including what may or may not be preventing you from discussing sex with your adolescent.

Reversing the Trend

Being a sex-savvy parent involves much more than parental controls on the internet or tuning out sex scenes on Netflix. Being the change you want to see in your teen begins with understanding your own sexuality and what you are communicating to your teen in your own behavior at home, as well as sex talks with your teens. What's required of you for modeling healthy sexuality is delving deeply into how your own early attachment relationships affected you. Your attachment, bonding, and intimacy capacity largely determine how your early years formed your sexual identity today and how it's nonverbally absorbed by your tweens, teens, and young adults. Being a sex-savvy parent is about you coming to terms with your own values about sex and intimacy. Is your sexuality objectifying, relational, or such a taboo subject you can't face it at all? If your sense of sexuality is not relational, how can you change yourself to become more relational?

In his book *Healing the Wounds of Sex Addiction,* Mark R. Laaser identifies boundaries, rules, roles, and addictive family behaviors that potentially influence the likelihood of teens developing an objectifying rather than relational view of sexuality. Boundaries can be either too rigid or too loose. Boundary violations can be as subtle as a mother leaning emotionally on her teen sons for what is better coming from her adult partner or, if not partnered, from adult friends. Domestic violence is another major boundary violation. Parents overreaction when discovering children masturbating or peeking on siblings naked are other boundary violations. More than once, an adult patient has disclosed to me that their mother found them masturbating and reacted by shaming them or reacting in other horrifying ways such as burning their fingertips while telling them masturbating is an unpardonable sin. If you think your family is affected by any of these signs of family dysfunction, your teen may be at risk for sex-related problems. It's vital to get professional help for yourself and your teen.

As parents, when we learn about our own childhood, bonding attachment needs, or injuries, we have a second chance to earn secure attachment for ourselves and in relationship with our teens as well. We can rewire our brains to earn secure attachment feelings. As Dan Siegel writes, "Neural integration creates coherence in mind and narrative; this can happen when we make sense of our lives, altering our attachment status and gaining 'earned secure' attachment." Earning secure attachment can occur somewhat in therapy, but its best completed more fully in the real life of mature parent-teen and adult partnerships. Most of my patients need both therapy and practice learning at home.

If we are going to be the change we want to see in our teens, it's critical that we parents make sense of our own early attachment relationships before we think we can begin to help our children with their attachment injuries in present or future love relationships. This includes how we express physical affection and our personal boundaries not only to our teens but to others in our family and community. It's also important to take a good, hard look at what adult relationships we have and if there are ways in which we look to our children to fill in our own attachment gaps. We need to dig deeper into our own difficult feelings related to relationships other than the ones with our children. As parents, our own relationships model and directly impact how well our teens can incorporate healthy sexuality as they transition into adulthood.

Responding More, Reacting Less

The toughest part of parenting is dealing with how we were parented and becoming responders rather than reactors to our teen's episodes of raging hormones. Pearson and others posed a question in their study related to how parents' own attachment gaps influence their children's ability to earn secure attachment relationships in adulthood. These researchers considered how variation in security in adult attachments related to parents' reports of depression and their own parenting behaviors. Participants were sorted into one of the four self-described attachment classifications—secure, anxious-avoidant, anxious-ambivalent, or disorganized attachment—and compared to their self-reported depressive symptoms. These researchers were interested in identifying connections between parents' own early bonds and their ability to engender secure attachments with their children.

Dan Siegel grapples with this very real problem with what he refers to as "parenting from the inside out." Questioning how your attachment bonds with your own parents/caregivers and reflecting on how these attachments have influenced your own parenting practices is vital. Even if your early attachments were not at all ideal, learning about earned attachment can help you break the negative intergenerational cycle you may be headed toward with your teen. A good start is questioning how you and your primary caregiver interacted with the following:

- eye contact
- facial expressions
- tone of voice
- posture
- gestures and touch
- timing of signals
- intensity of signals

Now take these basic body language signs and notice how you interact with your teen. What do your responses reveal to you about what you are conveying nonverbally in terms of your attachment approach capacity with your teen?

Relational vs. Objectifying Sex

The big irony of middle and late adolescence is that the body may appear all grown up and sexually prepared, but the brain has some catching up to do. Even when parents studiously talk with their teens, giving them the best preparatory guidance and cautionary tales about sex, if parents aren't savvy about how to have and hold this conversation, their words may fall flat to no effect. Becoming the change that you want to see in your teen's sexual identity requires your own understanding of how developmentally adolescents reach physical maturity five to seven years before their prefrontal lobes can grasp the consequences of high-risk sexual thoughts, feelings, and behaviors. Doctors agree that when parents talk to their children about sex, in many cases, these talks do make a positive difference in risk behaviors for their teens. For instance, when parents do discuss sex with their teens, teens are more cautious about the act of sex itself.

However, relationship intimacy calls to us in a much deeper way than just being sex smart does. If we want to become the change we'd like to see in our teens, delving deeply ourselves into our own intimacy and relationships and actually parenting from the inside out is the best way to become sex savvy for our teens. As most adults in relationships know, sex is far from just a physical act, and physical affection is far from being the only way we truly, deeply connect with those we love. So being the change, becoming sex savvy, is a whole lot more than just having "that talk." But beginning the dialogue about sex is an excellent start.

Have That Sex Talk

When it comes to misinformation, myths, and lies about discussing sex with our teens, polls of parents on the subject come up with the following:

- Talking to my teen about sex is like giving him/her the green light to start having sex.
- When my teen wants to talk about sex, he/she will ask me.
- It's better to scare them straight about sex rather than dialogue and interact about how sex is important in the context of relationships.
- What teens don't know (presumably from Mom and Dad about sex) won't hurt them.
- Tweens and teens are too young to discuss sex, and by the time they're young adults they already know about sex, so what's the point?

These may have been the myths that kept your parents from talking to you about sex but don't fall for any one of them. The wrong sex messages are all around your children. Be the change and give them the right messages about relational sex.

When parents lack comfort and clear communication skills about sexuality, adolescents are at the mercy of their peer groups and media preferences—and rarely do those sources help teens sort out and decipher the pain and pleasures or give deeper understanding of what intimacy and sharing with another human being is all about. Knowing how trivialized sex is portrayed in the media, how idolized teen sex is among peers, why do so many parents fail to talk about sex with their adolescents?

Brain Launch: Becoming Sex Savvy

What unhealthy habit will you drop, and what brain-healthy tools will you pick up?

1. By diving deeper into understanding yourself and your own sexuality, explore how sex happened for you as a teen and how sexuality is or is not healthy for you now in your adulthood.
2. Set the example for good self-care by reducing and managing your own relationship stressors in healthy ways so that teens find their own healthy coping for managing those raging hormones.
3. Identify your own early attachment relationships before you attempt to earn secure attachments with your teen.
4. Explore how you verbally and nonverbally express physical affection and where your personal boundaries are with not only your teen but also with other adults.
5. Stop expecting your teen to approach you about sex. Make time for a two-way dialogue about sex and intimacy.
6. Whatever you chat about, lecture less and listen more as much as possible with your teen. As a therapist, I see the communication train wreck that happens too often when a parent talks over their teen and then the teen shuts down or bolts out of the office with lightning speed—and at least for today, the potential

for connection and intimacy is gone in a puff of parental pontificating.

7. Take courage as a parent to share with your teen what your sex and intimacy experiences were like for you when you were a teen (while you zip your lip on oversharing intimate details). Share with your heart while allowing even more space for your teen to disclose what it's like for them now.
8. Rather than lecturing, consider sharing yours and mine notes. Your love note back to your teen is that more than anything, you hope their experiences are richer and more fulfilling than yours, if yours lacked. And you hope their experiences are even more awesome, if yours was just short of story book. Either way, being sex savvy for your teen involves creating moments of healthy emotional intimacy. This is the greatest stuff of life—don't blow it by lecturing.

Chapter 11

Love Repurposed

For this is how much God loved the world—he gave his one and only, unique Son as a gift. So now everyone who believes in him will never perish but experience everlasting life.

—John 3:16 TPT

Kurt's Brain on Love

Kurt is a 28-year-old who was diagnosed with oppositional-defiant disorder as a tween and now suffers from posttraumatic stress disorder (PTSD) as an adult. Kurt's biological mother drank while she was pregnant with him, which was the major contributor to his behavioral and academic problems. Kurt also had a considerable amount of childhood trauma even after he was adopted by a family member. The great-aunt and great-uncle who assumed custody of Kurt fought viciously. Their screaming matches and knock-down, drag-out fights continued through most of Kurt's childhood. Kurt's adoptive father had a temper that became even more volatile when he drank, and Kurt's father drank a lot. With all the developmental trauma Kurt had already experienced when he turned 18, Kurt was in for even more trauma as he left for active combat.

What's remarkable about Kurt is that after getting brain healthy along with weekly eye movement desensitization and reprocessing (EMDR) treating all that trauma, Kurt has now learned to cope with his PTSD much better. He has more patience with his children now that he has renewed his childhood faith in Christ and is brain healthy.

Brain Health Essential #8: Love Repurposed

Being the love that repurposes your parenting starts with seeking to understand your own love needs and clearing your line of sight to see past your needs to recognize your teen's love needs. As you're absorbing all the brain health principles we've covered so far, none of them will matter much unless principle #8—love repurposed—becomes your greatest ambition for your teen. But as Jesus said, our highest and greatest use of our brains is loving others *the same way* we love ourselves. The question then becomes, Do you love yourself? If not, are you able to love your teen despite what has gotten in your way?

The Brain's Love Connection

Oxytocin is a powerful hormone that acts as a neurotransmitter that accentuates feelings of a warm embrace, eye contact, or a deeper sense of belonging to one another when people are sharing a meal together. Oxytocin helps create bonding and attachment, about which Daniel J. Siegel, a child psychiatrist, writes: "Being around a secure attachment figure calms our inner turmoil and gives us a feeling of being 'home' and at ease." Oxytocin is the Cupid's arrow to our brain going straight into the hearts of those in close, caring proximity to us, making them feel as one with us.

That's its gentler side. Oxytocin is not all rainbows and pixie dust; it has a serious side. Dr. Siegel also warns: "[B]eware that this intensification [brought on by oxytocin] can be, especially for males, an intensification of jealousy and aggression...sexual involvement can create an intensification of romantic dopamine-driven obsession." The kind of bonding that is jealous and aggressive can cause teens to form attachments and bonding outside of the family circle that are premature and can disrupt identity formation. The single most important way to ensure that teens do not form unhealthy bonds outside the family is maintaining healthy bonds within the family. Most parents don't have to record a home video to remember some of the worst exchanges with their teens. I wish that I had learned about fair fighting with teens before my oldest daughter became a teenager. Joshua Coleman, author of *When Parents Hurt*, has advice that would have helped me a lot to avoid the emotional trap that parents and teens get into. Not getting hooked into an out-of-control anger fest with teens is one of the most important relational traps to stay clear of. Coleman first advises that when emotions are at their highest, it's not the time to problem solve. Here is a brain principle about stopping before conversations become knock-down, drag-out fights.

Within your teen's more potentially reasonable prefrontal cortex but above your teen's least affectionate and survivalist primitive brain lies the blessing and the curse of the relationship with your teen—the excitable, reactionary, and super-sensitive teen hippocampus. The hippocampus is relatively small when you consider how big its job really is.

Like the US Postal Service, the hippocampus has been called a "beehive" of the brain's short-term memory sorting system, with thousands of messages sent through it each minute from all over our sensory world. The hippocampus determines where all those memory "letters" should or should not be stored. At any point in time, without a hippocampus doing its sorting job, we might remember everything that happened before our hippocampus stopped sorting memory but all the memories thereafter will be gone, blowing in the wind. People who lack a working hippocampus experience life like *Groundhog Day* and *50 First Dates* all rolled into one. Without a hippocampus, the brain can't tell the difference between what must be remembered and what's best forgotten because it's all tossed out with every new day.

The hippocampus, also called the "seat of memory," oversees and records all kinds of memories from mundane to divine, from what you had for breakfast to meeting the love of your life for the first time. Right next to the hippocampus, also in the limbic system, is the amygdala (recall it's the brain's fight, flight, or freeze alarm), and as these two small structures meet at a crossroads of the limbic system, this whole emotional center is highly susceptible to hormones like adrenaline that are ready at a moment's notice to disengage or, if deemed necessary, fight. Putting it all together, cross the excitability of the hippocampus (which is feeding into the immaturity of the teen's amygdala) with your teen's loosely connected frontal lobes and you have a reasonable explanation for the highly charged emotions that can interfere in the parent-teen relationship. There is one more important element. It's bringing your teen's emotional system back down to a calm state after a high state of emotional arousal. That's another feature of immature brains—they can escalate quickly, but they don't always de-escalate as quickly.

When You and Your Teen Escalate

One of the parenting "snakes" I learned to pick up by the tail was knowing when to back off in an argument with my teen. This is what I wish I had recognized most when my oldest daughter and I would begin with a potentially heated conversation. "Don't take the bait" is what I tell myself now. It's so tempting to become defensive when it seems teens are purposely provoking with criticism, rejection, and fury all aimed at you. Of course, you are the totally innocent, even angelic parent—right? Even if you are not at all in the wrong and you truly are trying to help your teen, after he/she has put you on blast, respond by saying your "I" message peacefully and go no further if you see nonverbal signs of teen overload. These signs might be like head in hands, gripping fists, tightening facial muscles, or slumping down and blushing red. Give it up and call a time-out now—don't follow your teen down that emotional rabbit hole. Stay grounded and calm.

If possible, start the conversation again with empathetic bonding language that includes all of you: your body stance, facial expressions, and voice tone. But first take a break yourself, and give your overactivated brain time and space to find peace and calm. Then approach your teen with only bonding words and loving nonverbal communication. Be specific and genuine. Tell him or her in detail what you love about them, how you care, and how proud you are of specific accomplishments he/she has already made on whatever subject has become such a hot topic. Don't delve too far into details again; remember, the teen brain easily gets lost in details and feels as if you're blaming them.

Test this bonding approach first by using your senses to attune to your teen's experience. If it appears your approach did not produce the calming, bonding effect you were hoping for, check your body language and tone of voice. Internally stand looking in front of a mirror of yourself. How congruent were your words and actions? Did your words say, "Bond with me, I care," while your body said, "Don't just sit there like a hopeless lump. Just do what I say—now why won't you listen to me"? If you were standing there like Hercules, as if holding up the doorframe, while you caught your teen sitting slouched at his/her computer, what does this say to your teen about the power differential between you? What does this Herculean stance say about how much you want a collaborative friendly talk versus another totalitarian lecture?

Perhaps you are not holding up the house like Hercules, but your arms are folded tightly across your chest? Do your subtler signs of incongruent communication betray you, like sighing quietly in frustration or hanging your head down as you shake it in disbelief when your teen is trying to talk to you? If you cannot manage to prevent these feelings of your own frustration from bubbling to the surface, it's okay—just say so! To become the change, you want to see in your teen—become authentic, open, honest, and willing to share.

There is one important caveat: make every effort to keep your focus lovingly on your teen and off your own unmet needs—be fully present with your teen. Most communication about your teen isn't for resolving your own psychotherapy problems—that's what professionals like me are for. If you don't have a therapist and find yourself using your teen for one, figure out another way to get your emotional needs met.

Building Capacity for Bonding

Laurel Parnell writes that nonverbal attunement, also called limbic or empathetic resonance, between two people is nurtured through physical proximity, eye contact, facial expressions, voice tone, gestures, movements, undivided/undistracted time, and touch. Nonverbal attunement builds capacity for bonding with your teen and is largely accomplished through your body language while you're with your teen. Although it may feel like scripted behavior at first, your brain registers nonverbal attunement more naturally by relational parts of the brain.

One aspect of attunement is the imitative resonance of mirror neurons, which were first discovered in the premotor, supplementary motor, and inferior parietal cortex. Although scientists agree we have them and know how they work, there is a disagreement surrounding the purpose of mirror neurons. Are they clearly for mothers, fathers, and babies to bond in caregiving? Are they a protective survival mechanism assessing nonverbal signs of surprise attack? Are they truly part of our prosocial interaction with each other? Do mirror neurons help us build capacity for bonding and for developing a greater capacity for empathy in the human brain? To me, the answer is obviously yes. However, doctors and their complicated brains do like to debate such obvious things—don't they?

Mirror neurons fire when a person or animal sees the action of another and involuntarily copies the same behavior—for instance, tears welling up in your eyes as tears are streaming down a movie actor's cheeks. Maybe you can't help but chuckle and smile even if you're not completely sure if a comedian's punch line is funny. You laugh anyway because the audience all around you is laughing wildly. Scientists generally agree that mirror neurons are part of our empathic capacity to trigger either feelings of trust or mistrust. Certainly, when teens notice a discord between a parent's actions and words, though they may not consciously know why, they do have ample reason to mistrust what you are attempting to convey if your words and actions aren't congruent.

When your words and body language don't match for your teen, which one should they trust? I'm not sure about you, but I would trust the nonverbal first—wouldn't you? So, you said with an angry tone, stern eye contact, and crossed arms something like "I love you!" Your teen's mirror neurons probably ratted you out before the words shot out of your mouth. Although your words and body language disagreed, how you said "I love you!" really meant the opposite to your teen. What you really said was, "You disappoint me. I am angry with you." It's hard to hear "I love you" and expect your child to melt into empathetic attunement with you when your body language is screaming so loudly with incongruent misattunement. That's one major way you may be creating a trust-mistrust conflict in your teen when you talk. How mirror neurons help us interpret and attune may be the first step you can take toward connecting more lovingly with your teen.

Your mirror neurons also interpret your teen's attunement with you. If you are fully present, attending to what your senses say about your teen's experience, and reflecting back what he/she is saying rather than lecturing, all this congruent body-verbal language leads to bonding with your teen. As you are attuning to your teen's needs, you are building trust.

Bypassing Your Teen?

Spiritual bypass is a term coined by psychologist John Welwood that refers to a tendency to use spiritual or religious ideas and practices to avoid confronting unresolved emotional issues, psychological wounds, or unfinished developmental tasks. If you tend to spiritually bypass your own pain, you may be subconsciously expecting the same from your teen. For you as a parent, this might take the form of pushing your pastor or priest in front of your teen rather than working through difficulties as a means to deepen your relationships with your teen. It may be that you use Bible verses to override your teen expressing needs about something you view as too sinful to talk about. It may be that your teen can never truly measure up to your religious standards because he or she doesn't, at the moment, share your tenets of faith. The teen years are a time for teens to challenge the faith of their childhood, and superimposing your faith on him or her will only drive you two apart. Be on watch for committing a spiritual or religious bypass if you are prone to excessive perfectionism and goodness; tend to avoid and repress undesirable painful emotions; fear intimacy, closeness, or vulnerability with your teen, and/or avoid your own responsibility or accountability in relationship to your teen. If this applies to you, you may be conveying to your teen that your spiritual and/or religious values are more important than attuning with your teen and have become a major barrier to the parent-child relationship.

Psychologist Joshua Coleman encourages a gentle persistence. Don't give up too soon. If your teen is rejecting you, self-soothe and choose not to take it personally. When your teen is conveying hate of you, as Jesus said, turn the other cheek. Look to the cross, imagine putting that impression of hate into a container, label it temporary, and put it aside as you focus on your teen. Then, as you've turned the other cheek, think, "Learn to love instead." Your teen is learning to love and be loved. Taking personally and reacting defensively from what your mirror neurons tell you is a rejection from your teen will not build your teen's trust in you. As we've seen from previous chapters, your teen's brain is processing a lot of emotional stimuli and grappling with serious data crunching between those ears using a mishmash of mature and immature parts. At one moment he/she may hate you and at another long for your connection and comfort.

Parents, keep yourselves in check when a heated argument starts to raise your emotional temperature. Coleman says, "Try hard for a long time to have productive conversations," and know that sometimes it's not the time or place for a pragmatic, useful negotiation or, perhaps for the moment, any interaction with your teen at all. Sometimes they're just too tired to hear. Some things are much better after a nap or a good night's sleep, or after one of the other six lifestyle health practices is attended to first. You can always reconvene on this debacle at a potentially more fruitful time. Then you can try a talk again, but attune with your teen first. More important than winning an argument—or even negotiating an agreement—is nurturing a loving relationship first through attunement. Over time, bonding capacity grows between you and your teen.

Another way of attuning to your teen is appreciating their dreams and aspirations. Take those "crazy" dreams your teen has about the future (like becoming a lizard breeder, for instance). Don't do a Godzilla squash step on those dreams. Dreams like that are rarely what they end up doing. Crazy teen dreams become points of entry for action, encourage their exploration, cultivate their creativity, bring them along with you to see all the skills you have developed, and help them reach out and become more skilled than you at many things. When they connect with their dreams, teens are flapping the baby feathers off so their adult wings can bear their weight in flight. It all started with those "crazy" teen dreams.

Getting Your Home Back

There is one more important balancing point about repurposing parental love. That is loving your teen with golden rule love, loving your neighbor *as you love yourself* love. Part of loving yourself is getting ownership of your home back. Yes, you share your home with your teen, but if your name is on the lease or the title of your home and not your teen's name, it is, in fact, your home. I say this because in coaching parents and teens with some failure to launch problem, I often get the impression it's the other way around. The teen owns the home and the parent gets to live there emotionally and physically. These teens have parents convinced that if the parents make one false move, they'll be given a 72-hour notice to vacate.

Repurposing parental love also means loving yourself enough to set limits on their demands of your money, time, and energy. Not taking back your life and your home from demanding teens hurts not only you; it's hurting your teen. That includes giving in to guilt when your teen wants you to spend your money on what's not brain healthy, everything from chips and sodas to more expensive gaming paraphernalia. It's time to step back and take a good look at what your teen's healthy needs are and what your teen's unhealthy wants are—there is quite a big brain health difference between wants and needs.

Brain Launch: Love Repurposed

What unhealthy habit will you drop, and what brain-healthy tools will you pick up?

1. If you find yourself guilt-tripping to get your needs met with your teen or in some other way slipping into "it's an all about you" kind of conversation, consider whether that's a weight your teen is mature enough to carry for you.
2. Be tougher, thicker skinned, and bigger hearted with your teen. A favorite pastor of mine when I was a young adult used to say, "Be more like an elephant. Have a thick skin and a heart even bigger." Don't take your teen's negativity so personally.
3. In heated moments with your teen, study the difference between being teen-brain sensitive and your own mature adult brain being too emotionally reactive.
4. Keep a teen-brain-informed stand. If you have to, whisper to yourself a mantra: "At this time in teen brain development, my teen doesn't have all his/her wires connected yet." Say it again and again like you are crazy in love with the wonders of the developing teen brain.
5. Don't forget your teen's excitable hippocampus, immature amygdala, and those loosely connected frontal lobe wires that are reason enough for your teen's momentary highly charged emotions. Use active ignoring or affectionate distancing when your teen hits you where it hurts.

6. Keep yourself in check, and when a heated argument starts to rise, don't take the bait. Try hard to have a productive conversation, but know when to adjourn. Sometimes it's not the time or place for a pragmatic, useful negotiation or a productive chat at all—even if you know absolutely you are right.
7. If your teen says you talk too much or lecture too long, you probably do. If your teen says you interrupt them too much, you probably do. Even if you don't lecture much at all and you never interrupt them midsentence, let them finish what they want to say.
8. Stop multitasking while listening to your teen; give them your undivided attention. If you tend to treat conversations with them as if it's a chore or obligation, you're probably hurting their feelings without realizing.
9. Kindergartener rules for crossing the street apply to love repurposed: stop, look, and listen to your teen, and then proceed carefully. Be careful not to trip on too many words.
10. Consistently and routinely build capacity for attuning, attachment, and bonding with your teen. Take off your parent hat occasionally and reach into your authentic self. Whenever possible, steal moments to sit face-to-face, and as you're doing so, keep your focus off yourself and on your teen. As you're facing your child, look eye-to-eye, breathe calmly and deeply, and afterward embrace them for at least 20 seconds. If this kind of attunement does not seem possible right now for you and your teen because there's too much anger between you, the next chapter will help you bridge that gap.

Chapter 12
Your Relational ACE

As children develop, their brains "mirror" their parents'...own growth and development, or lack of those, impact the child's brain. As parents become more aware and emotionally healthy, their children reap the rewards and move toward health as well.
—**Daniel J. Siegel**, psychiatrist

Adverse Childhood Experiences (ACE) Questionnaire[1]

Consider the periods when you and your teen were each growing up, during your and your teen's first 18 years of life. If you answer yes for

1 "Relationship of Childhood Abuse and Household Dysfunction to Many of the Leading Causes of Death in Adults," published in the *American Journal of Preventive Medicine* in 1998, Volume 14, pages 245–258.

only yourself to any of these questions, give yourself a 1, and if it's true for both you and your teen give your relationship a 2.

1. Did a parent or other adult in the household often swear at you, insult you, put you down, humiliate you? Or act in a way that made you afraid that you might be physically hurt? If yes only for yourself, enter 1. If yes for you both, enter 2.
2. Did a parent or other adult in the household often push, grab, slap, or throw something at you? Or ever hit you so hard that you had marks or were injured? If yes, enter 1. If yes for you both, enter 2.
3. Did an adult or person at least five years older than you ever touch or fondle you or have you touch their body in a sexual way? Or try to or actually have oral, anal, or vaginal sex with you? If yes, enter 1. If yes for you both, enter 2.
4. Did you often feel that no one in your family loved you or thought you were important or special? Or that your family didn't look out for each other, feel close to each other, or support each other? If yes, enter 1. If yes for you both, enter 2.
5. Did you often feel that you didn't have enough to eat, had to wear dirty clothes, and/or had no one to protect you? Or were your guardians too drunk or high to take care of you or take you to the doctor if you needed it? If yes, enter 1. If yes for you both, enter 2.
6. Were your parents ever separated or divorced? If yes, enter 1. If yes for you both, enter 2.
7. Was your mother or stepmother often pushed, grabbed, slapped, or the target something thrown at her? Or sometimes or often kicked, bitten, hit with a fist, or hit with something hard? Or ever repeatedly hit over at least a few minutes or threatened with a gun or knife? If yes, enter 1. If yes for you both, enter 2.

8. Did you live with anyone who was a problem drinker or alcoholic or who used street drugs? If yes, enter 1. If yes for you both, enter 2.
9. Was a household member depressed or mentally ill? Or did a household member attempt suicide? If yes, enter 1. If yes for you both, enter 2.
10. Did a household member go to prison? If yes, enter 1. If yes for you both, enter 2.

Now add up your answers: _______

That's your parent-child relational ACE score. How do your teen's adverse experiences compare to yours? Do you and your teen have similar scores with similar adverse experiences? Or are they different? Is there a misalignment between you two or childhood experiences you both share? Your relational ACE score is only informational and does not determine at all how good a parent you are. The value of your relational ACE is to help you recognize how your own adverse experiences have impacted your parent-child relationship and areas where you may want to consider finding your own individual healing. The way you can begin is identify whatever 1s you may have and use those questions to further explore how you have taken that experience and translated it into your own view on parenting.

Your relational ACE gives a kind of risk score to how tenuous the parent-teen relationship may be. Dan Siegel calls this "the neurobiology of 'we'" or "interpersonal neurobiology." To oversimplify a bit, parents' brains are wired to relate to their children and children's brains are wired to relate to their parents. Just as we can rewire our own faulty individual circuitry by healing our own childhood trauma, we can also rewire interactively the parent-child relationship. But when parents leave childhood trauma of their own unhealed, the burden of this pain can become a legacy passed on to their children. Trauma-driven parents can "react out" the intergenerational traumas on their teens, essentially robbing teens of their own adolescence. Without conscious awareness of the hidden relational ACE at work, teens may act out, fighting the vicarious entanglement that they feel when burdened with their parent's unacknowledged and unhealed ACE.

The number one barrier to changing your teen may not be entirely all about your teen's bad moods, your teen's faulty thinking, or your teen's irresponsible behavior, but much more about confronting your own self-doubt, guilt, and shame connected to your ACE that may be lurking below the surface—the fears, pain, and losses you felt long before becoming a parent yourself. Confronting your own ACE may be the deepest, hardest work you will ever have to face, but if you do, your heart will open to your teen in ways you'd never imagined possible. Repairing your relational ACE is no exact science—but the price to your children is high if you fail to face how your own childhood shapes your parenting relationship.

Relational ACE and Parental Shame

Sometimes a shame legacy is benign, even beneficial from parent to child. It might be as simple as a parent feeling shame for never getting a college degree and the relatively harmless responsibility falls on their child to pursue higher education—not such a bad intergenerational burden to carry from parent to child. As long as the adult-child was guilt-free in choosing his/her major that launched a desired career—no harm, no foul.

Parental shame is damaging when an intergenerational burden has more complicated and deeper-reaching tentacles attached. For instance, when a parent demands that their child graduate from a prestigious university with honors and launch a career in a field that the adult child has no passion for at all. Or when a parent withdraws all support contingent on the child taking over the family farm like the parent did, even though the parent had no desire to become a farmer like his father and left all his dreams behind. When a child is driven to soothe parental shame rather than to become the person that child wants to become—that's when parental shame becomes toxic.

When parental shame is lived out vicariously in the adult child, it can also cause lifelong damage to the parent-child relationship and is clearly unhealthy for the parent. Parental shame is profoundly different than guilt, and it's inaccurate to give it the adjective *parental*, because shame begins in a parent's childhood before becoming a parent. Since shame carries such strong potential to wound teens deeply, it's worth a review of the difference between parental shame and guilt.

Whether it's you or your teen who feels it, guilt can be healthy as a warning when we've done something wrong and need to make an amends for our behavior. For a parent, a guilty act might include going overboard with teasing or failing your teen in some way, like forgetting to attend a special event important to him/her. Guilt also has a healthy recovery time and says, "I did something wrong, I should rectify it, and repair what I can." Parental guilt permits the parent to maintain a sense that, as a mother or father, despite the mistake, he/she is still a good person.

A parent's shame is a world apart from guilt because of the irreparable damage shame can cause to the parent-teen relationship. Shame cuts deeply into the core of our person, and it goes viral in the parent-child relationship. It's a self-blaming personal affront from a parent's childhood that carries an internal message: "I am bad," instead of "I did something bad." Many parents who bring shame into their adult life dream of becoming parents just to escape the feelings of emptiness that shame carries. These are the parents who say they wanted children to love and finally feel loved unconditionally. While no parent wants to feel unloved by their child, the difference with a shame-based wish to have children is that these parents are attempting to repair the ways their own childhoods lacked love.

Coleman writes that shame-based parent-child relationships are rooted in the parent's own childhood trauma that "left us feeling disfigured, with feelings of defectiveness, unlovability, or fear…[and] parenting becomes more treacherous as children grow because their capacity to reject, shame, and humiliate the parent increases in weight and power as they get older." Colman continues that the price the parent-teen relationship pays when parents have failed to identify and work through their own shame carried over from childhood is great and disables the hope of a truly healthy parent-teen relationship as long as it remains hidden.

The takeaway here is that if you want your teen to change, make sure you have cleared away what is holding you back from feeling fully loved and accepted by God's grace yourself, and confront your own shame. Uncover whatever has created your own shame core and choose to become the change that you want to see in your teen. By letting go of your own shame rather than vicariously insisting that your teen somehow cover your shame for you, you will not only free yourself but also interrupt an intergenerational legacy in your generation for good.

Ways Shame Got Passed on to You

Getting to know how you were influenced by your own upbringing begins with going deeper than the following common problems many of us had with our own parents including:

- Perfectionistic or overly religious parents
- Chronically depressed or anxious parents
- Overly protective or possessive parents
- Neglectful parents
- Self-absorbed or narcissistic parents

- Critical, abusive, or rejecting parents
- Domestically violent parents
- Drug- or alcohol-abusing parents

Going deeper means looking in that rearview mirror just long enough that you can see where you've been—not because you are backing up, but so you can see what might be ahead for your teen if you repeat the same mistakes your parents did. Start with finding your ACE score; you might be surprised to know even a 1 or 2 can have profound effects on how you parent your own children.

Unburdening Parental Shame

In his book *Shame and Pride: Affect, Sex, and the Birth of the Self,* psychiatrist Donald Nathanson identifies the relational black box of parental shame, which, when unexamined, becomes the latent cause for parent-teen communication problems. When parental shame is triggered, parents rely on one of four default responses in coping with their shame: attack their teen, attack themselves, withdraw from themselves, or avoid their teen. As long as shame is lurking in a parent, one of these feelings, thoughts, or behaviors will emerge, driving an ever widening, unhealthy gap between parents and teens. As interpersonal neurobiology suggests, when your teen's shame is triggered, your teen will also react in one of these four ways. When you get to know your relational ACE, you can reduce the risk of triggering these shame reactions in both of you.

When one of these four reactions occurs in you—attacking your teen, attacking yourself, withdrawing from your teen, or withdrawing from yourself—stop and consider how this reaction is impacting your teen. What particular ACE of yours or your teen's may have been triggered? Are potent, irrational feelings coming up for you while talking to or thinking about something your teen said, did, or failed to do? See what memories pop up for you from your childhood. If you cannot recall anything but still feel a foreboding sense of shame, take note of what your amygdala (that fight, flight, or freeze part of your brain) is telling you to do in your body where shame usually betrays itself. If that doesn't jog your hippocampus (that memory keeper) to direct another part of your brain to memory recall, start a chain analysis, which is writing down sequentially everything you remember about the event in order to understand it better.

The most salient caveat here is that shame is hidden, so get it into and keep it in the sunlight of consciousness so you can overcome it and become the change you want to see in your teen. The very nature of shame is that it's a bad feeling about yourself that is so uncomfortably painful and awful to you that your brain worked very hard to protect you from your conscious awareness of it. And it's likely true, back when you were a child, that if your survival brain had not protected you from your shameful feelings, it might have overtaken you before you were ready to process through it. Now, as an adult parent, its time to pick up your shame by the tail and deal with it. This is not something you have to do alone. A qualified counselor can help so that you break the legacy of pain before your teen gets saddled with it.

Part of growing into better parenting is bringing what's hidden in you out into the light of your awareness. The most insidious and damaging way that shame impacts parenting is causing parents to feel so fundamentally flawed that they form a shame reaction that usually is not healthy for their teen. For instance, parents punish themselves or act overly pleasing to teens. Or parents withdraw helplessly from their teen when the teen needs a loving teaching moment. Instead of self-punishing or withdrawing, parents, reach out to other parents. Commit to self-care by nurturing yourself. It does your teen no good when you punish yourself when your teen is misbehaving. It's a well-worn but useful analogy: put your own oxygen mask on before you help your teen strap on theirs.

Ways of Healing Your ACE

Sebern Fisher says, "All psychotherapy, when it helps, must in some manner enhance the mind's capacity to regulate the brain. We are all too often arguing over what manner we use—psychodynamic, body-oriented, analytic, or cognitive-behavioral—when what we need to discuss is how we best help the brain to regain or gain its regulation." Neurotherapies focus first on calming and regulating the brain and then on understanding the workings of the mind to make changes. Like the brain looking at a mirror of itself, self-directed neurotherapies are the most direct route to enhance the mind's capacity to regulate the brain because they each bypass long narratives about problems and feelings and require no medication. Neurotherapies simply allow the brain to use its own sensory experience to directly recognize what it's doing and then respond and regulate itself accordingly. "The mirrors" that neurotherapies use vary but, in all cases, are data derived directly from brain activity detected from its own sensory system—visual, auditory, tactile, visceral—so that the brain can more directly evaluate and regulate itself.

Although there are a number of other therapies that in various ways enhance the mind's capacity to regulate the brain, I will highlight three different types of therapy that help integrate and regulate better thinking, emotions, and decision making. These therapies are eye movement desensitization and reprocessing (EMDR), electroencephalogram biofeedback (EEG neurofeedback), and Internal Family Systems (IFS) therapy. Although each of these techniques approaches the brain differently, what they all have in common is accessing brain processing as a felt sense in the body or on a computerized display, reflecting brain regulation back in some meaningful way and bringing that feedback loop to conscious awareness so that the brain can reprocess and improve overall regulation.

The reason these therapies work with the mind to regulate the brain is that they each, in various ways, actively focus on brain integration. As Daniel Siegel describes: "Well-being emerges when we create connections in our lives and help the brain achieve and maintain integration, a process by which separate elements are linked together into a working whole." I've chosen these brain-focused therapies because each contacts disparate parts, connects, and integrates the brain into a better working whole.

Eye Movement Desensitization and Reprocessing

EMDR is a breakthrough therapy with special capacity to overcome the effects of psychological trauma. It was developed by Francine Shapiro, an American psychologist, in the late 1980s. Initially, EMDR was used and studied as a therapy for posttraumatic stress disorder (PTSD), which was a relatively new diagnosis at the time. EMDR therapy is an eight-phase treatment that comprehensively identifies and addresses experiences that have overwhelmed the brain's natural resilience or coping capacity and have thereby generated traumatic symptoms and/or harmful coping strategies. Through EMDR therapy, patients can reprocess traumatic information until it is no longer psychologically disruptive.

During this procedure, patients tend to process the memory in a way that leads to a peaceful resolution. This often results in increased insight regarding both previously disturbing events and long-held negative thoughts about the self. For example, an assault victim may come to realize that he was not to blame for what happened and that the event is really over, and as a result, he can regain a general sense of safety in his world.

Since the development of EMDR therapy, many adaptations of the therapy have been established to address particular types of psychological problems, such as PTSD; however, all specialized applications rest on EMDR's basic protocols and concept of adaptive information processing. EMDR has also been used effectively in the treatment of children who have experienced trauma such as great loss or child abuse.

For EMDR information, contact:

EMDR International Association
5806 Mesa Drive, Suite 360
Austin, TX 78731
info@emdria.org

Parnell Institute for EMDR
900 Fifth Ave, Suite 203
San Rafael, CA 94901
Parnellemdr.com

EEG Neurofeedback

Also called EEG biofeedback, neurofeedback (NFB) is a type of biofeedback that uses real-time visual, auditory, and tactile feedback displays of brain activity—most commonly electroencephalography (EEG)—to teach self-regulation of brain function. Typically, sensors are placed on the scalp to measure activity, with measurements projected using video displays or sound. Neurofeedback offers an additional treatment option for people with eating disorders, addictions, mood disorders, anxiety, and attention deficit disorder.

NFB has been around since the 1960s, but it had not gained huge traction as a mainstream intervention for several reasons. First, the pharmaceutical industry holds the market on medication interventions.

Second, prior to the advent of the personal computer, NFB equipment was cost-prohibitive; it is still expensive, but at least now it is within reach of the average clinician to provide. Third, up until recently NFB brain "games" did not hold the entertainment value they do now. Finally, until recently the clinical applications were limited to certain conditions for which its application had been widely tested, such as ADD/ADHD, generalized anxiety, and seizure disorders.

There is now a growing body of research showing neurofeedback efficacy in treating these disorders. Although by itself NFB is not a panacea, when used to enhance lifestyle health, NFB helps to optimize brain function. This is particularly true when all the lifestyle brain health practices are in place and the person's brain nevertheless continue to dysregulate, being continually overstimulated when the need calls for calm/alertness and understimulated when the need calls for focused attention. The brain typically self-corrects, but when the brain does not, continual dysregulation can become a new normal. NFB helps train or retrain the brain to function optimally, the way it was designed.

For neurofeedback information contact:

Institute for Applied Neuroscience
90 Acton Circle
Candler, NC 28901
www.ian-asheveille.com

EEG Info
6400 Canoga Ave., Suite 210
Woodland Hills, CA 91367
eeginfo.com

Internal Family Systems Therapy

The IFS model represents a synthesis of two brain-friendly paradigms: systems thinking and the multiplicity of the mind. As with most great work, IFS was not a new discovery but a new, more comprehensive approach. As founder Richard Schwartz writes: "Many other theorists have described a similar inner phenomenon, beginning with Freud's id, ego, and superego, and more recently the object-relations conceptions of internal objects." What's different about IFS is that it is self-directed rather than therapist-directed. IFS address many parts of the brain conceptually through multiplicity of the mind theory, including how "parts" of ourselves function in relation to each other.

Dr. Schwartz continues: "I learned that parts take on common roles and common inner relationships. I also learned that these inner roles and relationships were not static and could be changed if one intervened carefully and respectfully." Since IFS patients explore what circumstances forced internal parts of their mind into extreme and sometimes destructive roles, IFS is a profound revelatory tool, allowing them to delve more deeply into parts of themselves that carry burdens of shame.

Often, releasing these shame burdens has freed my patients to confront head-on why the lifestyle brain health practices covered in this book are too difficult at first to start and maintain with regularity. For instance, clients may defuse trauma such as sexual, physical, and/or emotional abuse, or a person's family of origin values and interaction patterns create internal polarizations that escalate over time and are played out in barriers to lifestyle health. Emotional eating is a prime example. IFS patients explore extreme parts that prefer to "stuff junk food" to metaphorically "stuff distressing feelings," and these kinds of extreme parts are compassionately acknowledged for their earnest effort to help the self and honored with new, less-destructive roles that are healthier.

For most people in IFS, their parts tend to play certain roles that Dr. Schwartz has named "Managers," "Firefighters," and "Exiles," whose overall goals are to keep the person functional and safe. He describes the efforts of the manager role to "maintain control of their inner and outer environments by, for example, keeping them from getting too close or dependent on others, criticizing their appearance or performance to make them look or act better, and focusing on taking care of others' rather than their own needs." As a result of being hurt, humiliated, frightened, or shamed in the past, exiles serve the role of carrying "the emotions, memories, and sensations from those experiences" and are "locked in inner closets" or become "incarcerated parts" we keep hidden. When emotions flare up, particularly with adolescents, we more often see the third group of parts—what Dr. Schwartz calls "Firefighters"—which tend to impulsively distract from painful shaming emotions by distracting with extreme behavior such as binging on drugs, alcohol, food, or sexual promiscuity.

Like other neurotherapies, by differentiating parts and integrating through awareness, the core sense of self is strengthened with what Dr. Schwartz calls "crucial leadership qualities such as perspective, confidence, compassion, and acceptance." He calls this self-leadership, which means that instead of becoming overwhelmed by their emotions, self-led people hold their center stable and become instead the stable "I" through the storms of life. For many teens, transition age youth, and their parents, IFS offers a self-directed brain-based therapy that facilitates remarkable support to practice the brain health essentials outlined here. IFS patients trade feeling chronically inferior and critical of self and others for instead exploring what Dr. Schwartz calls the eight Cs—compassion, curiosity, calm, clarity, courage, connectedness, confidence, and creativity—which lead to finding deeper relationships with ourselves and others.

For IFS information, contact:
The Center for Self-Leadership
PO Box 3969
Oak Park, IL 60303
www.selfleadership.org

According to the Center on the Developing Child at Harvard University, having at least one stable, committed relationship with a supportive parent, caregiver, or other adult at home, school, or church or in a community group can tip the scales in your favor, no matter how high your relational ACE is. Additionally, if I have not beat this drum a thousand times I have not beat it enough: getting your brain healthy and supercharged on its own neuroplasticity can help you and your teen overcome any adversity if you just put your brain to it.

Starting with learning all about how your own ACE score affected you, getting help from a qualified professional, and then identifying how your ACE plays out in your parenting will clear your line of sight to becoming that one supportive parent, caregiver, or other adult who actually can counterbalance any of your teen's ACE scores. You can become the change you want to see in your teen by nurturing those counterbalancing factors not only in your teen but also in yourself. Be the source of that one supportive adult-child relationship to your teen by helping your teen:

- Build self-efficacy and perceived control with the eight brain health essentials.
- Strengthen your relational ACE by recognizing when you or your teen are reacting out of shame.

- Reach out for help with counseling and get involved with a church community that has recovery groups to get further help with your relational ACE issues.

So, even if your ACE score is daunting and you feel counterbalancing resilience just isn't within your reach, start with one of the core brain health strategies for yourself: exercising, eating better, sleeping more soundly, reducing damaging stress, getting in recovery for substance dependency, checking out intimacy and sexuality for yourself, and taking steps to reach out and love others and yourself. What we know about neuroplasticity and the brain is that you and your brain have limitless potential for resiliency, adaptability, and positive change, no matter how high your ACE score is. This is just another way of saying you and your brain are headed for miraculous healing.

If your teen resists every single attempt you make at getting his/her brain healthy, it's no small consolation prize that you get healthy yourself. In fact, you getting brain healthy will improve your teen's chances of following suit soon enough—it's one of the greatest win-win circumstances of parenthood. My own mother used to say, "If you as the parent are okay, your child will be okay, too." So, this time, you get to be the change you want to see in your teen. If you don't believe me—or my mother—then believe the doctors. That same Harvard resilience study we just talked about found that adults who strengthen their own lifestyle health actually model the same resilience that their children need to overcome any ACE score. Just by you getting to be a better you and modeling healthy behaviors for your teen, the gift of brain resilience keeps on giving to your teen, even positively improving your genetic code, which can pass on from this generation to generations to come.

Chapter 13

Confronting Tough Teen Problems

Jesus responded, what appears humanly impossible is more than possible with God. For God can do what man cannot.
—Luke 18:27 TPT

At this point you may be discouraged that despite all your best efforts your teen's problem is too tough to change. It's times like this when breaking these problems down into smaller, more manageable ones might help. Here are some situations other parents have faced and ways that we've broken them down together into easier-to-deal-with solutions.

My Teen Won't Give Up Soda or Junk Food!

Regarding the question of soda, one mother says: "My teen has been drinking sodas for years, and it doesn't seem to have affected him. What's the point in taking his sodas away now? He's almost 13, and he's been drinking sodas daily since he was eight years old." Here's what I generally say: in light of neuroplasticity, it's never too late to stop unhealthy behaviors and never too early to start healthier ones. My experience has been that at first teens addicted to sodas do get upset when you stop buying them. If you truly want to improve your teen's brain health, you will stop stocking the house with sodas. Parent, be empowered. You have the power of the wallet, and it's your privilege to spend money on this brain-healthy project in the ways you see fit. Save sodas as a treat when you eat out; just don't keep a supply in the house. If they shop with you, just roll on by the soda section. Eventually they'll get used to it.

If you or your spouse likes sodas, too, consider giving your brain a break. Few liquids are more harmful to the brain than soda. A 2012 Harvard Medical School study found that participants who ingested just one 12-ounce soda per day were at greater risk of heart disease, chronic obstructive pulmonary disease, and, of course, dental erosion. Ditching sugary and diet soda can reduce your risk of pancreatic cancer by a whopping 86 percent. Soda spikes blood sugar, causing a burst of insulin, which leads the liver immediately to convert the sugar into fat. Over time, drinking just one can of soda per day interferes with neurological processes: it reduces brain-derived neurotropic factor (BDNF). With inadequate levels of BDNF over even a short time, soda-soaked brains are more learning impaired and less able to lay down, store, and retrieve memories compared to brains not battling soda's glucose overload.

My Teen Just Won't Follow Through

Another worry parents often have is their child following through with these brain essentials while their child is not under their direct watch.

These parents are again thinking brain health progress is like that typical elevator: either up or down. These parents think that this teen brain launch is a 24/7/365 burden they must carry all on themselves, making sure their teen will do all of this. I'd like you to think of the eight brain health essentials as all beneficial, so wherever you start, you can't lose. Even if you only manage to improve your teen's brain health by a fraction, small improvements go a long way toward inoculating against mental illness. If your teen fails to take responsibility for brain health on his/her own, it's healthy for you to share in brain health by teaming up with your teen. Do it together. Make it fun. Even if at first they don't think of it that way, they will eventually, especially if they can catch you slacking off on your own brain health. Give them choices. Make it a challenge in whatever way appeals to him or her.

For example, if your teen braces against your efforts toward relationship building, throw your teen off balance by giving him more of something that he likes but is a brain healthy variation. For instance, instead of more video gaming, consider trying "brain training." Try an EEG neurofeedback gaming consultation. The games are fun now and almost competitive with anything your teen is playing. A neurofeedback therapist can help your brain health cause along by chatting with your teen about the benefits of brain health essentials while he's playing a fun neurofeedback game.

Most schools these days are conscious of brain health—not allowing soda on campus, serving fairly healthy meals, undertaking physical education—so that's a good start. Focus on picking up the slack at home. Create a brain-healthy environment as best you possibly can. At home, start with the easiest brain health practices and work your way toward the more difficult ones. Again, think of brain health essentials as a wheel of good fortune that you can start anywhere. You will see the benefits as you continue incorporating them all into your family's lifestyle health.

Sometimes bribery is not such a bad tactic if it's an incentive for practicing a healthier habit. Just don't keep unhealthy food or drinks in the house even if you want to splurge as a bribery. Minimize compromise on the three core brain health essentials: activity, nutrition, and sleep. No brain can do without those three core brain essentials and ever truly be healthy. Also use gaming time as another incentive for building healthier brain habits.

At first, brain health does take a commitment, but once you and your teen become healthier even with just a few brain essentials on board as a daily practice, I assure you the momentum will continue because you both will be feeling so much better.

I Raised Myself; I Don't Know How to Parent

I've had some parents nearly give up on their teen because they were not parented well themselves. They believe they are at a serious disadvantage, even disability, to repurpose their own parenting. I typically encourage parents this way: start by reading fewer parenting books espousing perfect techniques. If you love your teen, which I know you do, you have a lot of good intuition built in to cover anything you may think you're missing because of a lack of confidence about parenting. Think of working on your ACE issues for your own benefit to learn about yourself and how you tick, not as evidence that you can't be a good parent.

As we covered in the previous chapter, one thing to watch out for when you have lacked good parenting in your own childhood is the risk of wanting to live that perfect childhood vicariously through your teen as a measure of your worth. Wanting love from your teen during this stage of your child's life is a bad idea. But really at any age, you need to live your life so that your child is unencumbered to fully develop into his or her emerging identity. This is particularly true in your child's tween, teen, and transition age years.

Remember, these are years when he/she is not capable of giving you enough love to ever make up for the love you lacked in childhood. Thinking you can get that unconditional love from your teen is detrimental to you both. Trust your gut more. What do you think your teen needs from you? Focusing on your own brain health is never a waste of time. As your mood, thinking, and behavior get healthier, the answers will come to you more easily. Your own good self-care will help build your confidence in parenting.

I Can't Deal with My Teen's Bizarre Appearance

A common question that may seem unrelated to brain health but actually touches on a very important aspect of repurposing parenting for better brain health goes something like this: "My daughter wants her hair neon green, and I can't stand that! How can I get her to see my way of thinking?" This is usually how I address that kind of question: Mom, Dad, sorry, but I am not feeling it with this "problem" because in the whole scheme of things, it's not a problem at all. Don't sweat the small stuff. Remember, you want to win the war for your teen's brain and overall health—no need to win every little battle. Accepting her desire for green hair is another way of surprising her. She will likely not admit it, but relaxing your views on minor, superficial, and literally cosmetic ways of her individual expression will help you win the war for more important things, like her brain. I mean, what's more important, her hair, which will grow out in six months, or the battle for her brain, which means winning her a better life?

Sex Is Too Embarrassing to Talk About

A question many parents are afraid to ask has to do with the sex talk. It's something like this: "Won't it embarrass my teen if I bring up sex with him?" I generally refocus on this question as a time for parents to sweat this big stuff. No matter how awkward the conversation is, *have it.* Communicating about sex is way too big to leave to chance or less reliable influences. Mom, Dad, be brave. Have the talk and keep the dialogue going about sex, sexual identity, and intimacy within the context of a committed, satisfying, mutually beneficial relationship. Do you want to leave your own values about sex, love, and intimacy to strangers or your teen's peers? I think not.

My Teen Is Glued to That Phone

Screen time is such a difficult barrier to optimum brain health that I often get comments like this: "I tried that media fast you advocate. Thanks a lot, it did not work. In fact it was disastrous! I unplugged everything, I took away internet, TV, smartphones, etc., just like you said, because I wanted her to go to sleep before 10 p.m. and get up and be ready for school on time. It didn't work. I'm tired of trying. I give up. What now?" This is one of those Wonkavator problems. So this particular media fast went sideways this time. Buck up, don't give up, and keep trying. Keep setting limits. It takes 21 days to get a bad habit unhooked and at least 21 more days to start a new healthier habit to take its place. Remember, you still are the gatekeeper, purse keeper, and the one in your household whose executive functioning should be fully on line to know what your teen needs. Your teen is not in charge; you are. And if you feel you must give in or give up, check out your ACE score items and see if something way back when you were young made you want to people-please your teen.

My Teen Is Too Stubborn to Listen

A common question relates to strong-willed teens. These parents will say something like this: "From the minute my child was born, he was strong-willed. How can I be expected to repurpose my parenting, let alone his brain, when he doesn't give me even an ounce of the respect I deserve as a parent?" I generally troubleshoot that one this way. Whether or not your teen is overly independent and controlling or not, sometimes we need to let teens fail while landing softly now so they don't hit harder on something bigger later.

Keep your eye on what would hurt your teen too much in the long run to let slide by now. Sometimes this means coming alongside teens and helping them in tandem, doing more together with him. Parenting is not an exact science. Sometimes we discipline too much, and sometimes too little. No matter what, if your teen doesn't recognize it now, he will later, that your imprecise, imperfect, but persistently kind, firm parenting meant you cared for and love him. The Bible says that parental love covers a multitude of ways we miss the mark of perfection. Even if your teen is strong-willed, he or she will respect you as you stay the course with improving your brain health while you are also laying the foundation for your teen to improve his or hers.

I'm Torn between My New Spouse and My Teen

With so many blended families in our country, I often hear both teens and parents complain that a stepparent is a barrier to teen success. I have personal experience both as a child myself who had stepparents and for my own two oldest children, who had my current husband as a stepparent. Recently, I had a father state his problem this way, "My wife is wonderful to me, but we argue about what we should do about my teen. She [as a stepparent] thinks I'm too lenient; I think she is too strict. I don't like how she and my daughter fight. What can I do?" Here, there is not a one-size-fits-all solution to this or the many other challenges blended families confront.

For sure, stepparenting can be especially challenging when teens are involved. If your teen's stepparent is overly critical, lacks understanding of teens, or expects too much or too little from your teen, it's important that you intervene—but not in front of your teen. Start private conversations with your offending spouse, and be gentle but persistent.

I have had too many adult patients who felt betrayed and unprotected by a biological parent's failure to intervene with an abusive stepparent to advise otherwise. This is another one of those "big stuff" kinds of issues that tags you, as the biological parent, as the one most important person who can help your teen and your spouse reduce stress and misunderstanding in your home. You are the parent. Take the lead now. Don't wait until your marriage or your teen suffers anymore without your active support. Your sweat over this will be worth it.

Don't let your spouse "make" you choose your child over his or her love. That's unfair and if he or she cannot stand the fact that you will not place your child's needs second, you must draw a line in the sand and take your child's needs into consideration first. If you do not intervene, and the marriage doesn't last, it's your child, not that spouse, who will have ultimately suffered. If you do intervene and this marriage lasts, your entire family will be better for you taking the lead as you help truly blend the needs of the family. Even in private, don't tell your teen your marital problems, and be careful not to undermine your spouse. That sets up a triangle of dysfunctional communication in the family.

Coleman's strategy for communicating with spouse stepparents provides important steps to bridging the gap between stepparents and your child. In his book *When Parents Hurt*, he suggests taking your spouse to the side, not in front of your teen, and trying the following:

1. Don't escalate with your spouse.
2. Start your parenting conversation with a compliment and self-disclosure about your own struggle with parenting.
3. When you catch your spouse and your teen in a heated argument, it's okay to intervene sometimes with just calling a timeout (e.g., *That's enough*, or *Let's take a break*).

4. Later, when you calm down and are feeling closer to your spouse, make a request like, "I'd like you to be the bigger person here. [Name the teen in question] feels like you don't want him around. I'd like you to help me out by reaching out to him and making amends. I'd really appreciate that."
5. Finally, it's important to let your spouse know if you're considering divorcing over your spouse's treatment of your teen.

Coleman states that he has seen many blended families overcome stepparent problems, and when the non-offending parent takes a stand on behalf of their teen, their marriage and family life improve.

I Can't Afford Brain-Healthy Food

Nearly every patient I introduce brain health to will comment that they just can't afford healthy foods. They will also say that, even if they can afford it, they don't know where to get food that is so specific. What I think they are really saying is, *Why start now?* I am also on a food budget, and yes, sometimes the healthiest food can be the most expensive. The less commercialized the food, the better. Do an internet search for a list of what is imperative that you get organic and what you'd be wasting your money on paying for organic. Other ways of sticking to a budget are buying fewer prepackaged meals and eating out less. Again, don't give up. Focus on basic whole foods for your teen and save money by eliminating processed foods that are higher in sugar, salt, and bad fats, and you can't go wrong.

My Teen Self-Harms

Self-harm is a sign of emotional distress usually caused by deep feelings of shame. Most often self-harm is kept secret, and being found out only compounds the shame. Self-harm can become a vicious addictive cycle that locks teens into a pattern of deeper shame. When I work with patients who self-harm, I've found several reasons they start: they are part of a peer group that self-harms, it distracts them from unbearable feelings, they want to feel physical pain because their emotions are numb, or they are practicing what it would be like to attempt suicide. The most common form of self-harm is superficial cutting anywhere on the body, but others burn, bang their heads, pull out hair, or pick at wounds so they don't heal. Even though for many teens this is phase, self-harm is not a behavior you want to ignore, and it's best to get professional help.

My Teen Gets Violent

It's astounding to think in the United States alone, as of the writing of this book, more than 187,000 students have been exposed to gun violence at school since the Columbine High School massacre occurred on April 20, 1999. Although risk of homicide seems like an entirely different problem than suicide, unless your teen has a serious drug problem, neurocognitive, psychotic, or personality disorder, in several ways, homicidal risk is actually just the flip side of the same shame-based problem as suicide.

I have worked in state prisons and hospitals, talking at length with men and women who want parents to know that even the best parents cannot rely totally on a healthy, God-fearing home life to ward off the potential for their child to enter into a life of crime. I have counseled at length with "lifers-without"—meaning they have no chance of ever being released because of the seriousness of their crimes. My point? With school shootings becoming so common, it's our duty as parents to wake up and pay attention to our teen in ways we never dreamed possible when our parents were raising us. Homicide, as a form of revenge, at its very roots is a shame reaction. It's important that you discover your teen's typical shame reaction, does it involve acting outwardly or inwardly? Does your teen attack others, attack himself, withdraw from himself, or withdraw from himself?

Not all children need to see a counselor like me. If you and your child have a good relationship, solid supportive social network, and caring family, that often is more than enough to keep teens from acting out violently. Even if your child's ACE score is as low as 2 because of experiences as common as divorce and a parent who uses drugs or alcohol, just these two factors, depending on the severity of the experience, can negatively impact some children enough to trigger violent behavior. If you have any doubt at all about your teen's potential for violence, and he or she has access to anything they can use to harm him- or herself or others, lock it up. This includes your own medication, alcohol, guns, knives, or anything else potentially harmful until you can be certain your teen does not have a plan to act out with his or her feelings of revenge. Being certain means you get the professional opinion you need to determine what your teen may need to address, including suicidal or homicidal feelings.

I'm Afraid My Teen Might Commit Suicide

I have sat with grieving parents who have lost a child as young as 12 and as old as 30 to suicide, and these were good parents. It's frightening to think that we can do everything in our power we know is right for our children as parents, and suicide can still strike our family. There are several risk factors that make tweens, teens, and transition age youth more vulnerable to suicide than middle-aged adults. One is that teen neurotransmitters and hormones, along with loosely wired prefrontal lobes, cause emotional ups and downs, which trigger impulsive acts as serious as suicide. The other is how unfortunately easily accessible drugs and alcohol are to our children. The third is that our children are being bullied. Finally, the family unit is being stretched so thin that children are too often left to fend for themselves with little loving, bonded adult care. This is not an exhaustive list, of course, but when these four factors are attended to more carefully, the risk of suicide is reduced considerably.

One mother expressed her pain this way: "I am a teacher where my daughter went to elementary school. I had no idea how much she was being bullied. She was one of the few white students in a mostly Hispanic school. That, in combination with me being a teacher, I guess she just thought she couldn't tell me. Maybe it was because she was being threatened. She was only 12. If I could turn back time somehow, I would have done everything I could in my power to stop this. One day I came home and she had hung herself in her closet."

Up to 12 percent of teens experience anxiety disorders. Fewer than one in five receive any treatment—psychotherapy or medication. And suicide is the second leading cause of death in the United States (following unintentional injury) among individuals 15 to 34. Parents, first we need to "pray without ceasing" or, as the Passion translation says, "make your life a prayer" (1 Thess. 5:17). A close second is making space in our busy lives to make sure we use all our God-given senses to recognize what's happening to our children. We need to spend not just time but *quality* time watching and listening closely to our children.

Parents, please don't feel alone. There is a lot of support for you and your child in preventing suicide. Following are some important signs. Watch for any sudden change in behavior and mood, even if they are suddenly very happy (this can be a sign of relief that they have decided that suicide is a solution). Watch for copycat suicide if there have been suicides of famous people in the media or with friends at school. Check for any indications of intent, means, plan, or history of failed attempts—these all can be risk factors that your teen has thought about or is thinking about suicide.

One of the culprits behind younger children committing suicide even as young as five years of age is bullying. Sadly, according to the US

Department of Health and Human Services, 20 to 30 percent of bullied students never tell an adult, so be nosy and don't give up until you know for certain that your child is not a victim of bullying. If you suspect your child is being bullied or is engaging in bullying, check out StopBullying.gov. There is a lot of helpful information that will help determine what to do next if he or she won't talk to you at all about bullying.

If your teen is quite vocal about thoughts of suicide but has never actually made an attempt on his or her life—even if it clearly seems like an idle threat to manipulate you—get your son or daughter into counseling with a licensed counselor for your peace of mind. But be forewarned: when it comes to suicidal or homicidal risk, your pastor or pastoral counselors should not be your first line of defense unless they are licensed in your state as a qualified mental health professional. Pastors, like many physicians, with the exception of psychiatrists, don't get adequate training in grad school to diagnose, assess, and treat mental illnesses.

If in doubt, call 911 and get your child an emergency mental health evaluation, or bring your teen to your local hospital emergency department. Emergency medical staff do know what to do to assess and treat suicide. Getting a professional's opinion may help you discover secrets your teen has been hiding. Secrets like drug or alcohol use, promiscuous sex, or being victimized by bullies are too often behind teen's suicide attempts. Look for signs of distress now and in the future. Just because you take your teen to church and youth group weekly does not protect him or her from these risk factors. With all the lovingkindness you have within you, be nosy and even intrusive—before it's too late.

I Need More Help. What Do I Do?

I have lived and practiced for over 25 years in a rural area of my state. Parents who have been struggling to get support for their parenting have many more options than when I first graduated from university. If you need a coach to help you through your teen's problems, no matter where you live, you do have more options than you might think.

My first suggestion is taking a careful look at your own ACE score compared to your teen's ACE score. If either of you have a score of 2 or less, you're probably all right to look up a local child-adolescent therapist in your area, preferably one who focuses on some type of self-directed neurotherapy and brain/lifestyle health program for you and your teen. A good place to start is your own local community primary care. Integrated behavioral health specialists are popping up in primary care clinics all over the US now. For instance, I work two days a week at a rural community primary care center providing integrated behavioral health to patients in the same building where they see their primary care physician—that's a perfect way for you to get started with lifestyle brain health.

If you prefer someone to meet with you or your teen in person but are not able to find an available mental health professional that takes insurance in your area, consider joining online lifestyle health coaching or finding licensed therapist who does online coaching in lifestyle health.

If you or your teen have a mental health, substance use, or major medical problem that requires close medical supervision, it's best to opt for putting together a team yourself for your support. For more information, you can write to me at Beautiful Minds via our website www.OakhurstCACounseling.com.

Chapter 14

Dare You to Move

Becoming the Change You Want to See in Your Teen

It's impossible to enjoy anything when you are afraid of failing at it. But once you know with all your heart that you really do have what it takes, being a mom can be a lot more fun.

—Joyce Meyers

Let's be straight. I'll admit that, in my more dramatic moments, I've thought Mother's Day would be better named Doomsday. Okay, not every year, just the years before I confronted my not-always-good-at-all but mostly good-enough parenting. As a therapist, I hear mothers and fathers lament about how Mother's Day and Father's Day feels so horrid to them. A 45-year-old patient who has never married or had children recently said, "You know, if Mother's and Father's Days were meant to only celebrate good mothers and good fathers I'd have no issue. But in my case, my parents didn't deserve any accolades, so I don't give them any on either day."

These days can be torturous reminders that children don't always honor us the way we'd like. Some of us deserve a day of honor, some not so much, but it hurts either way. From where I come from as a therapist, everything I've discussed in this book works—I've seen it with my patients and children. But I'm a mother, too, and if I were sitting where you are right now, worried about my teen, I'd have my doubts.

If I were you, I'd have to be saying, "Something good has just got to stick with my teen, or I am done trying." And you are done with pleading, yelling, ranting, crying, cajoling, bribing, punishing, withholding, setting boundaries, and worrying. No matter what you do, sometimes teens won't break, won't bend, and won't budge. Tweens, teens, and transition age youth can be infuriating, stubborn, and ungrateful. They can sit on their hands in defiance just because while you're wringing yours just because they're heading for disaster and failing to launch.

When teens act out—or implode with emotion—it's exhausting, worrisome, and heartbreaking. More than once with each of my four teens, there was a time or two I wanted to pack my bags and leave—if not for good, at least until they grew up! I fantasized that maybe I'd join an old hippie commune for exhausted parents like me, especially when well-meaning friends said, "They'll grow out of it," or "If I were you, I'd kick them out of the house," or "It's just hormones." I knew that, at least for me as a mom, none of these comments were helpful, especially when I saw that often these were the same friends who didn't have much of a relationship to their adult children. I just had to believe there was a better way to approach my teen problem.

The Good Habit Loop

Every teen and every brain is different, so how do I know this will work for you or your teen? Well, I'm sure it won't if you don't dare to move yourself, and I'm sure it will if you take up that dare and start getting brain healthy. And then there's your teen who may have different problems; maybe you're even thinking her brain doesn't have this one last brain operating system that will cause her to habituate to new healthier brain habits. What makes me so sure that you're such a powerful parent that you can get your teen to get brain healthy?

So you don't start thinking that I'm pumping both of your brains full of deceptive hopes and futile inspiration, I will confess I've left out one part of your brain that holds the key to your teen's success. This part of the brain isn't all that smart, but it will make it happen for your teen: it's the reptilian brain. That homely looking blob, called the basal ganglia, sits on top of that even less attractive brain stem. These two brain structures are important to making the eight brain health essentials work for launching you and your teen into a state of optimum brain health.

So, in case you are doubting any of this is going to work for your teen, I'm leaving you with one last brain lesson. It's how everyone's brain on the inside seems to look like your teen does on the outside—lazy. You can use this part of the brain to your parental advantage. Depending on your tween's, teen's, or young adult's temperament, it's true that you have more or less influence over your child's behavior. I get all of that, but I assure you that if your teen is breathing, swallowing, has a pulse rate, and is conscious enough to blink his/her eyes, your teen has this part of the brain. This part of the brain works with the basal ganglia located in the cerebrum that functions in many ways like voluntary motor control. It also works with the brain stem for procedural learning relating to routine behaviors or habits. But before the brain stem and the basal ganglia can be successful in perpetuating your teen's healthier brain habits, the orbitofrontal cortex (OFC), that area of the brain that relays goal-directed action, needs to relax. What that means for us as parents is we need to become what we want to instill in our teens. If we want calm, thoughtful teens, we need to become calm, thoughtful parents as a daily habit.

As Christina Gremel, assistant professor of psychology at UC–San Diego, asserts, "Habit takes over when the OFC is quieted." How do we do this? Gremel says, "We need a balance between habitual and goal-directed actions." Translated for us parents: it's a matter of both methodically repeating healthy daily routines and also breaking up these routines into more flexible blasts of novelty. Break up the monotony of doing everything "right" all the time by disrupting perfection (how boring) with fun. That is, you basically maintain the healthy habits you've established but intermittently switch it up with whatever is fun for your teen. Don't let you or your teen's brain lose interest and develop tolerance against those activating "go" neurotransmitters like dopamine or epinephrine. Disrupting good habits on occasion with fun gives that spark of pleasurable feelings that tend to get lost when habits become dull and monotonous.

"Have fun together and thereby help to develop a relationship based on enjoyment, mutual respect, love and affection, mutual confidence and trust, and a feeling of belonging. Instead of talking to nag, scold, preach, and correct, utilize talking to maintain a friendly relationship. Speak to your child with the same respect and consideration that you would express to a good friend." (D. Goldman and M. Goldman, *The ABCs of Guiding the Child* (1967)

Forming good habits to last a lifetime harkens back to simple psychology 101, when we learned that intermittent reinforcement is the best way to maintain any new behavior. So, parents, stop being so serious all the time. Never breaking these brain health "rules" is much too boring for you and your teen's miraculous brains. Sure, persist with healthier habits often, but don't forget what all this brain health is for. Don't forget to sit back and enjoy time with your teen once in a while.

Enjoying time with your teen doesn't have to be expensive, unless your time and attention are too expensive to waste. You might have to squeeze into your teen's busy schedule instead of him into yours, but do it. More than any other parenting practice, nothing is healthier for your brain or your teen's brain than spending time without expecting anything from each other, sharing a few moments together doing nothing more spectacular than living, laughing, and loving one another. This might mean sitting down and playing a video game or watching a movie with your teen. This may mean getting out on a short adventure together. It may mean sharing a hobby together. Whatever it means, make it something your teen likes to do even if he's not used to sharing it with you.

When I Started Becoming Brain Launched

I know many parents find Mother's or Father's Day difficult, and it sometimes becomes a major source of resentment. Many parents of adult children complain that they don't get a phone call or that children still at home forget to bother wishing them a happy Mother's or Father's Day. Parents just feel forgotten and unappreciated on "their" days. There was one Mother's Day I will never forget, when self-pity and parenting hit me smack in the face. My children were all old enough, I thought, to honor me, but I did not really believe I was much of a mother, so all I asked was a clean house and to have all four of my children together. Instead I felt "all dressed up with nowhere to go." Not only did my kids not get together and not clean the house, but my youngest egged my car! It was pretty shocking. All it said to me was that I am a bad mother.

That really fed my self-pity. The fact is that, though I am imperfect, I am a good mother. My evidence is that all four of my children, by the grace of God, are doing remarkably well. None have drug or alcohol problems. All are either working or straight-A students. All are reasonably happy and well adjusted. People like them, and they like people. Jesus said that like a tree, we will know the character and quality of the person by their fruit. With a whole lot of grace, my fruit, my children are evidence that the tree, though imperfect, is bearing good fruit.

So, rather than taking my children's shenanigans so personally, I really needed to enjoy the humor. My youngest son egged my car thinking it would make me laugh. They didn't clean up the house because I didn't tell them until the last minute that I wanted that. My older children didn't visit because they had other family engagements to attend hours away. Self-pity would not allow myself to see those issues from my children's perspective. Maybe self-pity has grabbed hold of you, too, on Mother's or Father's Day? You may know that deeper, even more shamed-based parental feeling of not being appreciated the way you think you should be by your children.

But let's get back to my favorite subject: me! Poor mother me, seriously. I thought all I wanted for Mother's Day was reasonable. Just a clean house, a hug, and four (even half-hearted) little "Shucks, ma, I love you" messages—that would have done it for me, I thought. Worse, I actually thought I was entitled to this dream Mother's Day. I thought I deserved it even more than honoring my own beautiful, wise, loving, elderly mother, who was there all the time being my kid's mom when I was attending graduate school and working full-time.

That Mother's Day, my kids had very different ideas about what would be fun on Mother's Day for me. A decade later, I get it. I get what I put them through. I get the price they paid for my own failure to launch—which is a code phrase for me failing to grow up and out of self-pity myself. What matters in this little story of woe is what Dale Carnegie said: "Feeling sorry for yourself, and your present conditions, is not only a waste of energy, but the worst habit you could possibly have." For parents, feeling sorry for yourself will only drive your children away. Be the adult for the rest of your life. Call them, write them, do whatever you can to support and love them. Be the change you want to see in them.

What I needed to understand that Mother's Day was that I was still too prideful to really love my children the way they needed me to. Pride is quite different from humility. Pride was rooted in my parental shame. Humility is the gateway to experiencing God's unfailing grace and the answer to becoming a better parent.

For too many years, my pride kept me covering up my shame with anger, self-righteousness, and pride—the kind of pride that is even more toxic to the love and work of God's Holy Spirit in my life, and perhaps yours, too. That Mother's Day I found that pride and self-pity were blocking my access to the fullness of God's grace as a parent. It didn't happen overnight; it was a slow degeneration to a point of despair as I moved further and further away from a relationship with my Creator. Slowly my relationship with the Lord became last in line after children, family, husband, and career. All priorities were off after I stopped humbling myself before the throne of grace. James 4:5–6 says it best:

> *Does the scripture mean nothing to you that says, "The Spirit of God breathed into our hearts is a jealous Lover who intensely desires to have more and more of us"? But he continues to pour out more and more grace upon us. For it says,*
>
> *God resists you when you are proud*
> *But continually pours out grace when you are humble.*

So then, surrender to God. Stand up to the devil and resist him and he will turn and run away from you. Move your heart closer and closer to God, and he will come even closer to you. (tpt)

When I came to the end of my self-pity, desperate, alone, and ready to change, I had one most important thing left to work on: pride. The pride that kept me from humbly keeping God first before all other idols: career, money, husband, and even children.

A Christmas to Remember

Up until now, I haven't been entirely forthcoming about my relationship with my children who are nearly all grown and gone. First of all, most of what I outline in this book I did not know and therefore did not practice with all my children all the time when they were small. Second, with each of my four children there were periods of time I needed to choose whether I would force the issue or leave my children to discover the natural consequences of a bad decision or unhealthy habits. Third, with my oldest daughter, we had a long period of estrangement that felt like a lifetime for me. Finally, and most importantly, I have learned to never, never, never give up on my children or the relationship. Although I believe in clear boundaries and natural consequences, I do not believe in emotionally or spiritually abandoning my children at any age when they are not following what I think is the right path.

I am passionate about helping parents go the course with their children, because each child is very different and as parents we, too, are in various stages of development in our own lives. Long ago and far away we weren't always parents. We were and are people, many of us people with unresolved emotional scars, psychological wounds, and unfinished developmental tasks of our own to accept and resolve. The problem is that most of us become parents long before we even become aware we have anything at all to resolve. So we enter into parenting with our own painful childhoods, stressful couple relationships, and unfulfilled hopes and aspirations. Some of the hardest times to parent are when one of our children is at an age that we had difficulty ourselves. For me, it was when my children were 13—that's when my parents divorced.

Thankfully, there is a gentle, gracious learning curve through life as a parent. I have been blessed to witness the trials and triumphs of my patients, some of whom now have adult children and grandchildren, who are waiting for their adult children to reach out to them for holidays, birthdays, Father's Day, and Mother's Day. These patients are unknowingly allowing their own unresolved emotional issues, psychological wounds, and unfinished developmental tasks to continue creating barriers between themselves and their adult children. The single most important question I have asked myself when one of my tween, teens, or young adult children and I disagree is, *Would I rather be right or be in relationship with my child?*

Although the estrangement with my oldest daughter was heartbreaking for me, I learned through the experience to trust the Lord with my children. My children first belong to Him. As Kahlil Gibran writes,

> *You are the bows from which your Children as living arrows are sent forth.*
> *The archer sees the mark upon the path of the infinite,*
> *and He bends you with His might that His arrows may go swift and far.*
> *Let your bending in the Archer's hand be for gladness;*
> *For even as He loves the arrow that flies, so He loves also the bow that is stable.*

I am learning to see my children as arrows in the process of propelling into life, and with every challenge parenting presents, my job is being that stable bow from which they will fly. I am learning to guide them when they need it and get out of their way when their aim is straight and true on what they are to become. They don't have an obligation to make me whole or confirm that I am a worthy person. Neither can I make them whole. I am so glad that during the time my oldest daughter and I were estranged, I let go of demanding love from her and instead kept praying for her happiness regardless of what my role was in her life. Holiday after holiday went by, and I missed her so much, but I refused to let bitterness prevent me from loving her from afar.

Joy came not long ago when my younger daughter and I got to drive my oldest daughter home from her last out-of-state postgraduate commitment after becoming a doctor of veterinary medicine. I'll never forget packing up nearly everything she owned in her small pickup—dog, guitar, and the three of us. Bringing my daughter home those three days it took to drive from West Texas to the west coast of California, I am so grateful that I did not give up on our relationship back when we were so painfully estranged. At the same time I held one truth: I became willing to accept that even though I did reach out to her to tell her I loved her, that didn't earn for me the right to change her mind about our relationship. The lesson I want to pass on to you is that ultimately the Lord is in charge, and sometimes there is nothing in our power to change our relationship with our children, but everything is in His power and in His time. So, partner with Him on behalf of your children no matter what.

Weary parent, do not give up, whether it seems you are experiencing success or failure, pray. In every decision regarding your children, always ask, *Would I rather be right or be in relationship with my child?* And then, on top of that, pray. Although none of my children are mentally ill, the same principles apply with children who have a mental illness; they need our prayers, and they need space for them to make decisions, even if we're fairly sure those decisions are wrong.

This year, Christmas time was the first time in a decade that all four of my children were together under the same roof for a three-night sleepover. This year I realized that it is never the wrong decision to keep faith in what matters and hold to Christ's law of love as most important above my own needs for love and acceptance by my children. It's my hope that although I have confessed my imperfect parenting to you, you will be inspired to build on the best of what is contained in this book, because in all truth, none of us raise our children perfectly. We can only lean on one another and find peace and solace in the fact that God is in control, and we are not.

Dare You to Move Mom, Dad

The most important thing I want to pass on to you is that it was easy to complain about how my children weren't making me feel good about myself by "failing" to do for me what I thought I deserved, but much harder for me to confront what I had done or failed to do as a parent and, deeper still, how my parents had left some unmet needs in me. I needed to find acceptance and peace.

The year my oldest child graduated from high school, the theme song at her graduation ceremony was "Dare You to Move" by Switchfoot. The lyrics speak to how positive change begins by daring you to move even after failure and even in the face of fear. I learned that my children are watching, waiting, and daring me to make the first move to continue to reach out to them even if they seem too busy with their own lives to make time for me, to see them and love them both as they are and as the best version of who they want to become.

Your children are daring you, too, to become the change that all along you wanted to see in them. As I found forgiveness right where I fell, I hope you do, too. I fell right where I could be a better version of me for my children, my patients, and most importantly for my ever present, everlasting, and ever loving Abba Father. It was when I stopped seeing my children as an extension of myself, as some false barometer of my success, and started encouraging them to get happy and get on with living their life as a part of our family—that was the moment of truth when I started to become a much healthier mother, daughter, wife, doctor, psychotherapist, and Christian woman. I hope the trials and tribulations of parenting have that transformative power in your life, too.

When I started, as Daniel Siegel says, "parenting from the inside out" is when I began to see how the gift of parenting is one of the greatest growth opportunities of life. I still don't have it just right. I miss that precise bull's-eye for perfect mothering daily, but I'm getting there. Progress, not perfection.

My wish is that these brain essentials are helpful to you and your teen. Only you can know what the tension is for you, between you now and the better you that you will become with a healthier brain. Launching that better, healthier brain between your ears and nurturing that more intuitive secret brain around your gut will cause you to break your own limitless record of parental, personal, and professional achievements. All of this will not happen unless you keep your connection to the power you need to make better parenting happen.

As you tap into brain health, it is no exaggeration at all to say, "You ain't seen nothing yet, baby!" Sometime in the middle of getting you and your teen's brain healthy, putting Christ first, and living your life more fully, you will realize, looking back, how natural it was becoming the change that you wanted to see in your teen. Blessings, all you weary parents. The job before you is the single most important job on planet Earth, and you have the brainpower to accomplish the task ahead. So, go do it. Become the change you want to see in your tween, teen, and transition age youth.

About the Author

Dr. Nina Farley-Bates received a master's degree from University of California Berkeley in 1993 and a doctorate in integrated behavioral health from Arizona State University in 2016. Applying the latest discoveries in neuroscience and trauma-informed therapies, she helps her patients overcome trauma, addiction, and painful pasts by supporting their first line of defense against helplessness, hopelessness, and despair: healthy behavioral change targeted to improve emotional and cerebrovascular health.

It's true that for the average brain, with time and the dedication of a caring therapist, talk therapy alone can help greatly. Dr. Nina found that fear-driven brains, however, work constantly in protective mode and can be treatment resistant, making traditional approaches impotent. Dr. Nina has learned *brain-friendly* ways to reach those fear-driven brains. These therapies, including eye movement desensitization and reprocessing (EMDR), EEG neurofeedback, and Internal Family Systems (IFS) therapy, help people find their own remarkable capacities, strengths, and resiliencies. Each of these brain-focused therapies involve activating the brain's own neurogenesis by revitalizing the brain's natural, unlimited potential for creativity, growth, and healing. As the mind changes the brain, the brain can change everything in a human life—even fear-driven brains—for the better. Along with healthier brains through lifestyle health, these therapies improve the whole person—body, mind, and soul—in ways talk therapy and medication alone cannot.

Dr. Nina lives with her family in the Sierra Nevada foothills near Yosemite National Park.

Acknowledgments

To the Morgan James Publishing team: Special thanks to David Hancock, CEO and founder, for believing in me and my message. To my author relations manager, Margo Toulouse, thanks for making the process seamless and easy. Many more thanks to everyone else, but especially Jim Howard, Bethany Marshall, and Nickcole Watkins.

Thank You

Dear Reader,

One of my passions is keeping up to date on the most effective and least labor-intensive ways of getting brain healthy for people of all ages. Getting brain healthy and living from the overflow of God's grace doesn't have to be complicated. On the other hand, just reading an book about brain health won't get you in the middle of that abundance alone. God made us to love and share together.

As a thank-you for reading my e-book, I want to give you my ***Make My Brain More Miraculous*** coaching videos and **CBT-Lifestyle Brain Health Tracker** program, the same program I use with my own patients in my clinics. This free introduction program will help you get your son's or daughter's brain healthy while you get your brain healthier, too.

My challenge to you is this: when your brain starts working better, stop and notice how you feel about your relationship with your tween, teen, or transition age youth. I will bet that even if their problems have not changed dramatically, your ability to cope with your child will improve—miraculously.

When we love our brains the way God intended us to, your brain will love you back! I promise. Jesus said the fulfillment of all God's law starts with loving the Lord and also loving your neighbor as you love yourself. There is nothing more *you* than your brain. So, love *you* the way Christ commanded us to!

I invite you to love yourself enough to love your brain to health. If you want to receive my eight videos and the brain health tracker program, I want to send it to you at no cost because you deserve more ways to love yourself as Christ loves you.

Just email me through my website, http://www.OakhurstCACounseling.com. I get a lot of emails and don't want you to get lost—so in the subject line, please give me the book reader's code phrase "Make my brain more miraculous," and I'll make e-mailing you back a top priority.

Thank you!

Dr. Nina

CPSIA information can be obtained
at www.ICGtesting.com
Printed in the USA
JSHW021453230223
38138JS00001B/27